Rocket Science-101

by Patrick H. Stakem

(c) 2017

Number 12 in the Space series.

Table of Contents

Introduction..5
Author...5
Introduction..6
A note on Units..6
Rocket Scientist versus Engineer..7
    What is Rocket Science?...7
    What is Rocket Engineering?...7
What is STEM?..8
What is a Cubesat?...10
The Space environment..13
    Zero G issues 14
    Vacuum...14
    Thermal environment 14
    Orbital Debris...15
    Mechanical and Structural Issues...15
    ESD sensitivity...16
    Spacecraft Charging...16
    Radiation effects..17
Getting to Space...18
    TRL..20
    Rocket Engines..22
        Solid fuel..22
        Liquid fuel..23
        Ion propulsion..24
        Solar sailing...24
        Space elevators and mass drivers 24
    Orbit control 25
    Orienting in space – landmarks. 25
    Sun sensors..25
    Earth sensor...26
    Star sensor...26
    Magnetometers..27
    GPS..28
    Attitude Determination and Control 28
    Housekeeping tasks...29

    Consumables inventory and management..........................30
    Thermal management...30
    Electrical Power/energy management....................................31
    Antenna Pointing..31
    Radiation Damage Mitigation..31
    Safe Hold mode..32
    Flight software...32
  Orbit and Ephemeris...33
Celestial Mechanics...34
  Orbital Decay..35
Talking and Listening to a Satellite..35
Living and working in Space – the ISS...37
  How do I go to the bathroom?..38
  What time is it?...38
  How do I sleep?..38
  Can I go outside?..38
Should we be spending money in Space?..38
Getting fancy, multiple units..39
  Constellations...39
  Clusters...39
  Trains..40
Exploring our solar system – the complexity of the Problem...........40
  Exploring the Sun...42
  Exploring Mercury...44
  Exploring Venus...44
  Near Earth Objects...45
  Exploring the Asteroid Belt, dwarf planets, and Centaurs........45
  Exploring Comets...47
  Exploring Mars...47
  Exploring the Gas Giants...49
    Jupiter..50
    Saturn..51
    Uranus...53
    Neptune...53
    Exploring Pluto, and beyond..54
Manufacturing in Space...56
Space Tourism..57

Wrap-up..58
Bibliography..58
Resources..62
Glossary of terms..63
If you enjoyed this book, you might also be interested in some of these..76

Introduction

This book covers the topic of what is popularly referred to as Rocket Science, seen as a daunting topic, completely incomprehensible. This is targeted to the non-specialist as well. I am not a rocket scientist, but I know a lot of them. I are a Rocket Engineer. I'll explain the difference later.

This material will not prepare you for a career in Aerospace, but hopefully it will get your interest going, and explain some of the basics.

This book could be used in a STEM Program, but it is not targeted to that exclusively. It can perhaps help the STEM educator with a few key concepts, and provide some references and sources of information. The author does have a book published on STEM-Cubesats.

Author

Mr. Patrick H. Stakem has been fascinated by the space program since the Vanguard launches in 1957. He received a Bachelors degree in Electrical Engineering from Carnegie-Mellon University, and Masters Degrees in Physics and Compute Science from the Johns Hopkins University. At Carnegie, he worked with a group of undergraduate students to re-assemble, modify, and operate a surplus missile guidance computer, which was later donated to the Smithsonian. He was brought up in the mainframe era, and was taught to never trust a computer you could lift.

He began his career in Aerospace with Fairchild Industries on the ATS-6 (Applications Technology Satellite-6) program, a communication satellite that developed much of the technology for the TDRSS (Tracking and Data Relay Satellite System). He followed the ATS-6 Program through its operational phase, and worked on other projects at NASA's Goddard Space Flight Center including the Hubble Space Telescope, the International Ultraviolet Explorer (IUE), the Solar Maximum Mission (SMM), some of the

Landsat missions, and Shuttle. He was posted to NASA's Jet Propulsion Laboratory for Mars-Jupiter-Saturn (MJS-77), which later became the *Voyager* mission, and is still operating and returning data from outside the solar system at this writing. He initiated and lead the international Flight Linux Project for NASA's Earth Sciences Technology Office. He is the recipient of the Shuttle Program Manager's Commendation Award, and has completed 42 NASA Certification courses. He has two NASA Group Achievement Awards, and the Apollo-Soyuz Test Program Award.

Mr. Stakem has been affiliated with the Whiting School of Engineering of the Johns Hopkins University since 2007, and Capitol Technology University. Mr. Stakem supported the Summer Engineering Bootcamp Projects at Goddard Space Flight Center for 2 years He developed and taught a series of Cubesat Courses at the college level. Most recently, he has been helping STEM Projects across the globe introduce space topics into their curriculum.

Introduction

This book will try to explain in simple terms about space, space flight, humans in space, and a few more top level topics.

A note on Units

I am fairly conversant in both English and Metric units (what is the metric equivalent of furlongs per fortnight?). Metric (SI) is mandated for NASA usage now, for interchangeability with our partner space faring nations. When a lot of the legacy flights discussed here were flown, English units were the norm. I have tried to keep the units comparable to the mission at the time. Conversions are easy enough, but units conversion is a source of error. It's not what you know about units and measurement, its how you think. And, I still think English units (even the English use Metric now), and convert in my head or on my phone.

For scientific/engineering work, the Metric system is well thought out. For artisans, the English system served well, as most units were divided by 2. Which is easy. Fold the cloth. Hopefully, when we are all taught Metric first, some one will still remember the conversions. You just need a slide rule....

Ok, bottom line: for International ventures, the majority rules, and the use of the Metric System will be strictly enforced. NASA has embraced this rule.

Rocket Scientist versus Engineer

At first, we define a scientist as someone who wants to acquire new knowledge. An Engineer is skilled at building things. That's a huge simplification. Most scientists like to "get their hands dirty" building special purpose instrumentation and test gear. For an engineer, getting to build and program something is what they live for. We might generalize that the scientist asks "what" and the engineer figures out "how." And, the fields are not mutually exclusive.

What is Rocket Science?

This is Science related to rockets and outer space. Rockets are what we currently use to put people and stuff into orbit. It's not the only way to do it. With lower gravity, like on the Moon, we could build a space elevator. More about that later. The science part is about exploring and finding out what's going on outside our atmosphere. That's a lot going one that we need to know about. This ranges from, "will I need an umbrella tomorrow?" to "when will that big asteroid hit the Earth?" It is obvious that the dinosaurs did not have an effective asteroid avoidance program. And look a what happened to them. Basic Knowledge, Knowledge for Knowledge's sake. That is what drives science. Needing to know more.

What is Rocket Engineering?

Engineering is a process that allows us to build, test, and use new things. It is sometimes driven by science, and sometimes by commercial interests. There are specialty areas in Engineering, such as electrical, mechanical, optical, and such. It takes a team with diverse skills to build and fly a satellite. Engineers like to get their hands dirty.

What is STEM?

STEM (Science, Technology, Engineering, Mathematics) is the key to the United States' continued dominance in High Technology. It took a lot of expertise to implement the first cell phone. Now they are turned out like cookies in third world countries.

STEM addresses overall education policy and curriculum sources in schools, at critical grade levels.

Although the teachers are experts in their particular area, and know how to present grade-appropriate material, they may not know how to find and access access the resources they need, or where to get help in a particular topic area.

STEM programs are seen as critically important in the education system, world-wide. It is getting to be a complex, interconnected ecosystem. Advances in the subject areas of STEM will take place only if you know how to exploit this ecosystem for knowledge.

When I was in school well before the Internet and STEM age, I had an encyclopedia. Today, students can access WikiPedia from their smart phones. The focus has changed from knowing facts, which are at your fingertips on demand, to leveraging facts to innovate. This approach touches all of the academic disciplines, the Humanities, Languages, Art, besides the STEM topics. Perhaps the best skill set to have is good internet search skills. Teachers have had to transition from asking factual questions, to asking questions that derive from applications of online research, and accrued knowledge.

When I was a kid, there was no STEM. My interests in science and engineering led to research and hands on experimentation. Luckily,

I survived. I was called on, while in grade school, to lecture and demonstrate some concepts of electricity to a High School class. The first satellite was launched, and I was glued to the black & white TV. I participated in Model Rockety at the High School Level, and went on to fly Nationally. This was made possible by an extraordinary High School Science teacher. I made quite a few friends, some of whom became Astronauts. I was given a great opportunity when I received a full scholarship to a College of my choice. I went to Carnegie Tech in Pittsburgh (now, Carnegie-Mellon University), and launched a career in Engineering and Aerospace. It is time for me to pay forward.

I think that Aerospace should be a major focal point for STEM, embracing a wide variety of topics at the cutting edge of technology and science. I have a handful of technical degrees, and spent 42 years at the various NASA Centers. I teach Electrical Engineering courses world-wide, and have done specialty Cubesat courses at the undergraduate and graduate level. It is time to apply that expertise earlier in the education process.

My thesis is, a project brings together all of the interesting pieces to provide a focal point for student work. There is a massive body of applicable free support material available from NASA and the Aerospace community. This involves the Education offices at the various NASA Centers, the Visitor's Centers, the speakers' bureaus. I have taken on the task of making STEM educators aware of this vast treasure-trove of resources. I have experience teaching Cubesat engineering and operations at the advanced undergraduate and graduate level, but I have no experience or credentials at the critical pre-K thru 12 levels. Well, I had my grandson build a a solar system model when he was 3. He did a nice job.

I think STEM is a critical resource for understanding and implementing the future. I think the Cubesat paradigm is a good thing to introduce into STEM. Let's do this. Future generations of STEM-mies will thank us. Possibly from the Mars-base.

What is a Cubesat?

A Cubesat is a small, affordable satellite that can be developed and launched by college, high schools, and even individuals. The specifications were developed by Academia in 1999. The basic structure is a 10 centimeter cube, (volume of 1 liter) weighing less than 1.33 kilograms. This allows multiples of these standardized packages to be launched as secondary payloads on other missions. A Cubesat dispenser has been developed, the Poly-PicoSat Orbital Deployer, P-POD, that holds multiple Cubesats and dispenses them on orbit. They can also be launched from the Space Station, via a custom airlock. ESA, the United States, and Russia provide launch services. The Cubesat origin lies with Prof. Twiggs of Stanford University and was proposed as a vehicle to support hands-on university-level space education and opportunities for low-cost space access. This was at a presentation at the University Space Systems Symposium in Hawaii in November of 1999.

Cubesats began as teaching tools, and remain in that role, although their vast numbers in orbit showed they they have become mainstream.

In what has been called the Revolution of smallsats, Cubesats lead the way. They represent paradigm shifts in developing space missions, opening the field from National efforts and large Aerospace contractors, to individuals and schools.

Even if your personal Cubesat project never gets launched or even built, it will bring you valuable experience to participate. This book will introduce and explain the NASA Systems Engineering Process, which leads from a set of goals to a successful space flight. This model has been in use for decades, and has proven itself to have usefulness. It has been refined as problems were uncovered, and still remains a viable approach to space missions, as well as regular engineering applications.

Cubesats can be custom made, but a major industry has evolved to supply components, including space computers. It allows for an off-the-shelf implementation, in addition to the custom build.

There is quite a bit of synergy between the Amateur Satellite (Amsat) folks and Cubesats. NASA supports the Cubesat program, holding design contests providing a free launch to worthy projects. Cubesats are being developed around the world, and several hundred have been launched.

Build costs can be lower than $10,000, with launch costs ranging around $100,000, a most cost-effective price for achieving orbit. The low orbits of the Cubesats insure eventual reentry into the atmosphere, so they do not contribute to the orbital debris problem.

Central to the Cubesat concept is the standardization of the interface between the launch vehicle and the spacecraft, which allows developers to pool together for launch and so reduce costs and increase opportunities. As a university-led initiative, Cubesat developers have advocated many cost-saving mechanisms, namely:

- A reduction in project management and quality assurance roles .
- Use of student labor with expert oversight to design, build and test key subsystems.
- Reliance on non-space-rated Commercial-Off-The-Shelf (COTS) components .
- Limited or no built-in redundancy (often compensated for by the parallel development of Cubesats) .
- Access to launch opportunities through standardized launch interfaces.
- Use of amateur communication frequency bands and support from amateur ground stations.
- Simplicity in design, architecture and objective .

Multiple cubesatas can be carried as secondary payloads on military and commercial flights, Cubesats, as small, inexpensive units have a higher mission risk tolerance.

Since the initial proposal of the concept, further efforts have been made to define internal and external interfaces made by various developers of Cubesat subsystems, products, and services that have defined the Cubesat 'standard' as it is today. A core strength of the Cubesat is its recognition of the need for flexibility in the definition of standards, and since conception the standard has evolved to ensure that these design rules are as open as possible. The most significant of these further advances in definition have been for the POD systems (in order to meet launch requirements) and the modularization of the internal electronics.

The in-orbit success rate of university-led Cubesat projects (not withstanding launch failures) is around 50%; this is an understandable result of using the Cubesat as an education tool, where development itself is a learning process and in-orbit failure is a disappointment but should not be considered the primary focus. For projects involving significant participation of companies with experience in satellite development, all but one were a success and demonstrated the strength of the Cubesat for non-educational applications. A large number of Cubesat missions have demonstrated significant success in-orbit operations for a sustained period. All Cubesats missions have had technological objectives to some degree, be it the demonstration of devices and system architectures developed in-house, or demonstration of Non-Space-Rated (NSR) Commercial-Off-The-Shelf (COTS) component performance

Earth imaging is a common objective for a Cubesat mission, typically achieved using a CMOS camera without any complex lens systems. As it is a critical impediment to the development of a highly capable platform for mission operations, the testing and evaluation of novel approaches for increasing downlink data rate and reliability is also a common objective. While less common than Earth imaging, real science objectives are becoming increasingly popular as recognition (primarily by NASA) of Cubesat capabilities increase and collaborations between engineering and science groups emerge.

Another important and related aspect in the design approach is that of modularity in a complete and integrated Cubesat life cycle, effectively representing a modular system of systems. The accelerated life cycle demonstrated consistently by small satellites, and harnessed by many Cubesat developers, can be further enhanced by the application of modularity to the complete life cycle. Cubesats are ideal teaching tools for aerospace engineering students, even if they are not going to fly.

Cubesats can fly alone, as secondary payloads with other missions such as the MARCO Project to Mars, and in Swarms. The MARCO mission has 2 Cubesat fly-alongs, that separate after launch, and continue to Mars along with the primary payload.

What type of missions do Cubesat's do? Initially, they served as communications relays for Amateur radio. But, they can do essentially what any "big" satellite can do.

The Space environment

We'll discuss the environment in which satellites operate, assuming they survive the launch.

The space environment is hostile and non-forgiving. Is there gravity in space. Of course. It is a relationship between two masses. I the orbital case, between the satellite and the Earth. It's just that the satellite is traveling very fast, and it balances out the gravitational pull. In fact, the satellite is in the gravity field of everything else in the solar system and universe. Most of that stuff is too far away to make much difference. But, recall, the Moon effects the oceans -we call them tides. With no gravity, no convection cooling is possible, leading to potential thermal problems. Space is a high radiation environment, being above the shielding provided by Earth's atmosphere, and the magnetic field.

There are differing environments by Mission type. For Near-Earth orbiters, there are the radiation problems of the Van Allen belts and the South Atlantic Anomaly, the thermal and vacuum environment, and the issue of atmospheric drag. This drag causes orbital decay, where the spacecraft slowly descends. There is also a drag factor from the residual atmosphere and the solar wind, and the spacecraft's orbit can be affected in other ways. All Cubesat missions are currently near Earth, although NASA is developing specialized Cubesats for planetary exploration.

Zero G issues

Zero gravity, actually, free-fall, brings with it problems. There is no convection cooling, as that relies on the different densities of warm and cool air. Any little pieces of conductive material will float around and short out critical circuitry at the worst possible time. And then, there are the strange issues.

The Hughes HS 601 series of communications spacecraft suffered a series of failures in 1992-1995 due to relays. In zero gravity, tin "whiskers" grew within the units, causing them to short. The control processors on six spacecraft were effected, with three mission failures because both primary and backup computers failed. This is now a well known materials issue, with recommendations for the proper solder to be used. In 1998, the on-orbit Galaxy IV satellite's main control computer failed due to tin whiskers.

Vacuum

The Satellites operate in vacuum. Not a perfect vacuum, but fairly close. This implies a few things. Lubricants evaporate and disappear. All materials outgas to some extent. All this stuff can find its way to condense on optical surfaces, solar arrays, and radiators.

Thermal environment

In space, things are either too hot or too cold. Cooling is by conduction to an outside surface, and then radiation to cold space. This requires heat-generating electronics to have a conductive path to a radiator. That makes board design and chip packaging complex and expensive. You get about 1 watt per square meter of sunlight in low Earth orbit. This will heat up the spacecraft, or you can convert it to electrical power with solar arrays.

Parts (and you) can be damaged by excessive heat, both ambient and self-generated. In a condition known as *thermal runaway*, an uncontrolled positive feedback situation is created, where overheating causes the part to further overheat, and fail faster.

There can be a large thermal gradient of hundreds of degrees across a satellite, where one side faces the sun, and the other side faces cold space. There is a similar situation at the planet Mercury, where one side always faces the Sun, and the other, deep space. It wiggles a little, creating what is called the "Goldilocks Zone," not too hot, not too cold.

Orbital Debris

There is a huge amount of debris in Earth orbit, including old booster rockets, failed satellites (Zombie-sats), broken solar panels, nuts and bolts, a Russian Space Suit. Space is large, but all of this stuff constitutes a hazard to ongoing missions. All this stuff is tracked and reported by the U.S. Air Force. There is a requirment now that old, end-of-life satellites have to reneter the atmosphere and burn up.

Mechanical and Structural Issues

In zero gravity, everything floats, whether you want it to or not. Floating conductive particles, bits of solder or bonding wire, can short out circuitry.

Another issue in vacuum is the cold-welding of certain metallic materials. This occurs when two pieces of material, without an oxide layer, are pressed together. This is facilitated by having very clean surfaces, and a vacuum environment. This affects moving subsystems such as solar arrays and steerable antennas. Early deployment of mechanisms is usually not a problem, but mechanisms that have to move throughout the mission, such as solar arrays and antennas, can be problematic.

The satellite structure has to take into account launch loads from the rocket, have sufficient stiffness and at the same time have low mass and be inexpensive.

ESD sensitivity

Solid state devices are particularly susceptible to electrostatic discharge (ESD) effects. These effects can involve very large voltages that cause device breakdown. Certain semiconductor lattice structures that have been damaged can actually "heal" over time, a process called annealing. Passive parts are sensitive to ESD as well. As parts are made smaller, the susceptibility to ESD effects increases. Proper grounding helps with ESD, providing a consistent voltage across components, without significant differences. ESD lead to sudden catastrophic failure.

Spacecraft Charging

Another problem with on-orbit spacecraft is that they are not "grounded." This can be a problem when a potential develops across the structure. Ideally, steps were taken to keep every surface linked, electrically. But, the changing phenomena has been the cause of spacecraft system failures. Where does the charge come from? Mostly, the Sun, in the forms of charged particles. This can cause surface charging, and even internal charging. Above about 90 kilometers in altitude, the spacecraft is in a plasma environment At low Earth orbit, there is a low energy but high density of the plasma. The plasma rotates with the Earth's magnetic field. The density is greater at the equator, and less at the magnetic poles. Generally, electrons with energies from 1-100 keV cause surface charging, and those over 100 keV can penetrate and cause internal charging. As modern electronics is very susceptible to electron damage, proper management of charging is needed at the design level.

Just flying along in orbit causes an electric field around the spacecraft, as any conductor traveling through a magnetic field does. If everything is at the same potential, we're good, but if there's a difference in potential, there can be electrostatic discharge. These discharges lead to electronics damage and failure, and can also cause physical damage to surfaces, due to arcing. This has been a problem at the International Space Station.

Radiation effects

On Earth, we are shielded from most radiation by our atmosphere and magnetic field. The Sun is the major source of our radiation, and high-energy particles. Right now, based on International Space Station experience, an Astronaut can have a maximum duration in space of about a year, before receiving his/her maximum lifetime dose. At the ISS you get, in a day, what some one on the ground would accumulate in a year. Shielding is one answer, but it can be counter-productive. Sometimes, a hit by a massively energenic particle can cause a spray of lower energy particles, from the shielding itself.

A large solar flare occurred in September of 1859, and was observed by British astronomer R. C. Carrington in his private observatory on his estate outside of London. Both the associated sunspots and the flare were visible to the naked eye. The resulting geomagnetic storm was recorded by a magnetograph in Britain as well. They also recorded a perturbation in the Earth's ionosphere, that we now know is caused by ionizing x-rays. In 1859, this was all observed, but not understood. Even the ionosphere was not know to exist at the time. Now, we know a Coronal Mass Ejection from the sun, associated with a solar storm, is first seen as an energy burst hitting the Earth, and later by vast streams of charged particles, that travel slower than the speed of light. At normal levels, these particles are seen as the Northern or Southern lights. The Earth's magnetic field is affected.

What did happen, and was not immediately associated with the solar storm, was interference with the early telegraph systems of the time. The telegraph was relatively new, and wires stretched for many miles. Think of them as long antennas. The telegraph equipment was damaged, and large arc's of electricity started fires and shocked operators. No fatalities were reported. The employees of American Telegraph Company in New York found they could transmit messages with the batteries of their systems disconnected. The Northern lights were visible from Cuba. This was the largest such solar flare in at least 500 years...and so far.

What if such a super flare occurred today? First, we would have warning from sentinel satellites such as the Solar Dynamics Observatory, that are closer to the sun, and detect the passage of particles. They can tell us about this via radio, which travels faster than the particles. So, we would have a day or so's notice. All of our modern high-technology infrastructure would be at risk of damage, from the electrical grid to the Internet. Most of our satellites would be damaged, removing services we rely on such as long distance data communication, and navigation. It would be much better to turn everything off, and ride out the storm. Even that might not prevent major damage to networks. When is the next large solar event? Even the Astrophysicists can't tell us that. Only that it will eventually occur. Stay tuned...

Getting to Space

Even without a launch vehicle failure, the launch environment represents the worst case in the path to orbit. There is significant shock and vibration, and a large acoustic environment. This only lasts for a few minutes, but is severe. The spacecraft must be tested beyond the expected launch environment, data for which is available from the launch vehicle provider. The Thermal/Vacuum and Vibration/Acoustics tests are usually referred to as "Shake and Bake."

Generally, we want to launch from a location near the equator, and on an Easterly course. This give us the maximum boost we can get from the rotation of the Earth. Satellites can also be launched into a polar orbit, but this takes more energy. The U.S. Launch sites are Cape Kennedy and the Wallops Flight Facility on the East coast, and Vandenberg Air Force Base in California, used for polar orbits.

There is also a floating launch facility in the Pacific. The Europeans use a launch site in French Guiana, on the equator in South America. The Russians use the vast deserts of Khazakstan.

It takes a huge expenditure of energy to get a satellite off the ground and up to orbit, moving around at 17,500 miles per hour to stay in a stable orbit. Well, that's what the launch vehicle, (or rocket), is for. Launch vehicles can be winged craft (like the Shuttle), and take off from an airport (like the Shuttle couldn't).

TRL

The Technology readiness level (TRL) is a measure of a device's maturity for use. There are different TRL definitions by different agencies (NASA, DoD, ESA, FAA, DOE, etc). TRL are based on a scale from 1 to 9 with 9 being the most mature technology. The use of TRLs enables consistent, uniform, discussions of technical maturity across different types of technology. We will discuss the NASA one here, which was the original definition from the 1980's.

Technology readiness levels in the National Aeronautics and Space Administration (NASA)

1. Basic principles observed and reported
This is the lowest "level" of technology maturation. At this level, scientific research begins to be translated into applied research and development.

2. Technology concept and/or application formulated
Once basic physical principles are observed, then at the next level of maturation, practical applications of those characteristics can be 'invented' or identified. At this level, the application is still speculative: there is not experimental proof or detailed analysis to support the conjecture.

3. Analytical and experimental critical function and/or characteristic proof of concept.

At this step in the maturation process, active research and development (R&D) is initiated. This must include both analytical studies to set the technology into an appropriate context and laboratory-based studies to physically validate that the analytical predictions are correct. These studies and experiments should constitute "proof-of-concept" validation of the applications/concepts formulated at TRL 2.
4. Component and/or breadboard validation in laboratory

environment.

Following successful "proof-of-concept" work, basic technological elements must be integrated to establish that the "pieces" will work together to achieve concept-enabling levels of performance for a component and/or breadboard. This validation must be devised to support the concept that was formulated earlier, and should also be consistent with the requirements of potential system applications. The validation is "low-fidelity" compared to the eventual system: it could be composed of ad hoc discrete components in a laboratory

TRL's can be applied to hardware or software, components, boxes, subsystems, or systems. Ultimately, we want the TRL level for the entire systems to be consistent with our flight requirements. Some components may have higher levels than needed.

5. Component and/or breadboard validation in relevant environment.

At this level, the fidelity of the component and/or breadboard being tested has to increase significantly. The basic technological elements must be integrated with reasonably realistic supporting elements so that the total applications (component-level, subsystem level, or system-level) can be tested in a 'simulated' or somewhat realistic environment.

6. System/subsystem model or prototype demonstration in a relevant environment (ground or space).

A major step in the level of fidelity of the technology demonstration follows the completion of TRL 5. At TRL 6, a representative model or prototype system or system - which would go well beyond ad hoc, 'patch-cord' or discrete component level breadboarding - would be tested in a relevant environment. At this level, if the only 'relevant environment' is the environment of space, then the model/prototype must be demonstrated in space.

7. System prototype demonstration in a space environment.

TRL 7 is a significant step beyond TRL 6, requiring an actual system prototype demonstration in a space environment. The prototype should be near or at the scale of the planned operational system and the demonstration must take place in space.

The TRL assessment allows us to consider the readiness and risk of our technology elements, and of the system.

Rocket Engines

This section discusses rocket engines that are used to get the payload to space, and to maneuver. What does the rocket engine push against in space? Wrong question. It doesn't push against anything. It is a reaction engine, using Newton's law of equal and opposite action and reaction. We need a fuel and an oxydizer. These can be liquid or solid, or a hybrid of both

Solid fuel

In a solid fuel, the fuel and oxidizer are mixed together in the correct proportions, and cast into a tube. There is usually a long void in the center, to expose more burning surface at ignition, and get a greater thrust for take-off. Once you light off the solids, you are on your way – there is not way to stop. Let's look at the Space Shuttle, which used both solid and liquid engines. The solid booster rockets were attached to the Shuttle's external tank. They did their job, were jettisoned, and recovered from the ocean to be re-used. The large external tank contained liquid hydrogen and liquid oxygen. The liquid rocket engines were located at the rear of the Shuttle orbiter. There were three, and they were steerable.

The launch sequence for the Shuttle proceeded this way: One by one, the liquid engines were ignited with a several second interval, and it was ascertained that they had reached their optimal thrust. If not, they could be shut down. With the solid engines bolted to the launch pad, and the liquid engines going, the tip of the orbiter rotated forward about a meter, but it wasn't going anywhere.

When everything looked good on the Shuttle liquid engines, the solid engines were ignited, and the explosive bolts holding the assembly to the pad were fired. Now you were on your way. During the couple of minutes of solid engine burn, you just sat back and enjoyed the ride. When the solids burned out, they were jettisoned. At this point, if there was a problem, the Shuttle could jettison the fuel tank, and do a return-to-launch site abort. Further along in the mission, out over the Atlantic, the Shuttle could do an abort to a contingency airfield. It there was a less serious problem there was an abort-once-around where the Shuttle could do one circumnavigation of the Earth, and land back at the runway at Kennedy Space Center.

Liquid fuel

Liquid fuel engines can be stopped and restarted, and can be throttled. The main problem was getting the liquids volatilized and into the rocket chamber. The combustion process creates large pressures, and this had to be overcome . Tanks pressurized with inert gases were used. The big breakthrough came for Werner von Braun working on the V-2 Rocket in Germany during World War Two. He implemented the turbopump, which is a small rocket engine using the same fuel and oxydizer, but turning a turbine. This turbine drove the fuel and oxydizer pumps. The ones on the first stage of the Saturn-V vehicle developed about 1,000,000 horsepower.

A hybrid rocket engine has a solid fuel, but a liquid oxydizer. It is partially throttle-able, by controlling the flow of the oxydizer.

Specific Impulse is the figure of merit for a rocket engine, and is in units of seconds. What? Wait, it's not "time." Ok, hold on a second (no pun intended). Specific impulse measues the total change in momentum. Momentum is mass times velocity. It is calculated by dividing the thrust (in units of force) by the mass, and you wind up with the units of "seconds."

Ion propulsion

Aonther propulsion system that works in space, but won't get you off a planet is ion propulsion. First we need large solar panels. We use the captured electrical energy to accelerate ions of a gas to very high velocities. The thrust is small, but can be sustained for long periods of time. The engines are super efficient.

Solar sailing

A solar sail is made of a lightweight reflective material. It uses the Solar wind for propulsion, in the same way a sailboat uses its sail and the wind.

In 2015, the Planetary Society's LightSail-1 successfully deployed its solar sail. This was done in Earth orbit. Planned follow-on projects LightSail-2, 3, and 4 will follow. Lightsail-2 will be launched shortly after this book goes to press, with 32 square meters of sail, and advanced guidance electronics. It is a 3U Cubesat. It will deploy its sail at 800 km. Further missions are also planned.

If you're really clever, you can make your light sail out of flexible solar cells, and use the collected energy to operate you Ion engine.

Space elevators and mass drivers

Two inexpensive options, compared to rocket launches, to get lunar material to orbit are the space elevator, and the mass driver. The space elevator is a concept that would work nicely on the moon. We have the advantage of a lower gravity than Earth, no atmosphere, and it could be built with currently available materials. There is nothing nearby to interfere with its operation. The lunar elevator would span about 50,000 km. The elevator needs a solid tether on the surface, a large mass at the upper end, for a tether, and a very strong cable. A lunar elevator could be tethered to a mass at the L1 Lagrange point, between the moon and the Earth. An elevator on the back side is also feasible. Space elevators have been explored since the 1890's. We now have the technology to construct them. A handy asteroid could be used as the counterweight for the lunar elevator.

A mass driver is an electromagnetic catapult, utilizing a long, linear motor. This works well on the lunar surface, with its reduced gravity compared to Earth, and lack of atmospheric drag. The other advantage is 15 days of sunlight, to operate the driver as well as charge batteries. In a mass driver, the payload does not contact the launch rail, but is magnetically levitated.

Orbit control

Because planets are not uniform in composition, neither is their gravity field. There are bumps and dips. And, the satellite is also gravitationally affected by everything else in the solar system (the whole Universe, actually). So, periodically, we need to adjust the orbit. This can be done with thrusters, or, if the planet has a magnetic field, we can push against that with electromagnets, called torquer bars.

Orienting in space – landmarks.

Getting ourselves oriented in space, in Earth orbit is relatively easy. See the big, bright thing? That's the Sun. See the huge circle – that's the Earth. Look north. That's Polaris, the north star. We can also have a Star catalog stored in the onboard computer, with several thousand of the brightest stars, That provides a great navigation aid.

Sun sensors

Sun sensors are universally applied to spacecraft. At least around Earth, the Sun is significantly bright to distinguish it from any other celestial object. Out in the outer planets, the Sun is less distinctive. Knowing where the Sun is located relative to the spacecraft is essential for proper orientation of the solar panels, to ensure a flow of charging current to the batteries. It is also a good rough orientation and orbital posiioning tool. Eclipses of the sun with respect to the spacecraft can be confusing unless anticipated.

Earth sensor

An Earth sensor, or horizon sensor, can see the interface between cool space, and the warm Earth. This is usually done in the infrared spectrum. Generically, you can use a horizon sensor for a spacecraft in orbit around any planet. Horizon sensors can sometimes be disrupted by the Sun, if it is in the field of view. The relative size of the planetary images give a rough value of the satellite's altitude. For a satellite in Earth orbit, if the moon is visible to the side of the Earth, it interferes with the image. The same is true around other planets with moons.

Star sensor

Star sensors measure star coordinates in a spacecraft frame of reference and provide attitude information when these observed coordinates are compared with known star directions obtained from a star catalog (data base). Star sensors can achieve accuracies in the arc-second range. Most star sensors consist of a Sunshade, an optical system, an image definition device which defines the region of the field of view that is visible to the detector, the detector and an electronic assembly. The detector such as a photomultiplier transforms the optical signal into an electrical signal. Solid-state detectors may be noisier than photomultipliers. The electronics assembly amplifies and filters the electrical signal from the detector. If the amplified optical signal from the detector is above a fixed signal intensity, an output is generated signifying the star's presence.

A charge transfer device star sensor is an optical system consisting of a digitally scanned array of photosensitive elements whose output is fed to an embedded microprocessor. A charge pattern corresponding to the received image of the star field viewed is produced, and is stored in memory for later processing.

The star sensor data is applied to a table of known stars. These "fixed" stars can be used for position and attitude information. Brightness is also used to verify the star identity. The star table is generally in celestial coordinates, and the observation is made from a known body-fixed position of the star sensor on the spacecraft.

Magnetometers

Magnetometers measure the induced current in a coil by a planet's magnetic field. This works well at Earth, where the magnetic field has been well mapped. For planets with little or no magnetic field, such as Mars, magnetometers are not very useful. A more sensitive measurement is made with a fluxgate. A fluxgate magnetometer consists of a small, magnetically susceptible core wrapped by two coils of wire. An alternating electric current is passed through one coil, driving the core through an alternating cycle of magnetic saturation. This constantly changing field induces an electric current in the second coil, and this output current is measured by a detector. In a magnetically neutral background, the input and output currents will match. When the core is exposed to a background magnetic field, it will be more easily saturated in alignment with that field and less easily saturated in opposition to it. The induced output current, will be out of step with the input current. The extent to which this happens depends on the strength of the magnetic field.

GPS

The GPS in-orbit satellite-based navigation can be used by satellites below the GPS spacecraft orbit (at 12,600 miles) for time and position services. Generally, four GPS satellites must be in view simultaneously. Commercial off-the-shelf rad-hard GPS products for spacecraft are available. This, of course, assumes the planet you are in orbit around has a constellation of GPS satellites. Only Earth, so far.

Attitude Determination and Control

To discuss attitude determination and control, we need to delve into Physics, discussing coordinate systems, kinematics, and dynamics.

The attitude of a spacecraft, regardless of the size, can be determined by inertial platforms, and with reference to external bodies such as the Sun, the Earth, the Moon, or the stars. For Cubesats in low Earth orbit, the in-place GPS system can be used. Besides our onboard orbital model, we might need models for the Sun and moon positions, a star map, and a magnetic field map.

The gravity field of the Earth is not at all uniform, and this effects satellites in free fall. In addition, the solar wind provides a bias. The drag of the Earth's atmosphere, even at orbital distances, also tends to force the Cubesat to slow down, and eventually re-enter.

Our points of reference include the sun and the Earth. The sun is easily seen (if it is visible form our position – the brightest light in the sky. Of course, it's not much help at "night" when the Earth comes between the Cubesat and the Sun. But, there's good old Earth. It's big, and its warm, compared to space. An "Earth sensor" can scan and see the transitions from warm to cold. For attitude control in the body frame, we'll probably use inertial sensors – MEMS gyros. Every once and a while, we have to adjust the gyros by referencing a known position, like the Sun. For a less exact attitude with respect to the Earth, we can use a 3-axis magnetometer, and a magnetic field model, stored onboard.

The attitude determination and control is done by the flight computer, using the various sensors for input, and the thtrusters for control. A lot of the time, the spacecraft will have momentum wheels. If you spin these in a particular direction, the spacecraft spins in the opposite direction, at a rate of the ratio's of ther masses. The wheels don't weigh much, but can be spun at high rates.

Attitude control can take the form of reaction wheels, where a small mass is spun rapidly, and the large structure of the spacecraft moves in the opposite direction, preserving momentum. Reaction wheels use electrical motors. One problem occurs when there are biases that cause the reaction wheel to saturate (achieve maximum speed) in one direction. It must then be unloaded of momentum by reaction jet firing (involving electrically controlled valves), or torquer bars, which, when energized, push against the planet's magnetic field. Electric propulsion, or pulsed plasma propulsion can also be used to make attitude and orbit correction.

Gas jets can use a cold-gas source such as nitrogen, or a hot gas system with a propellant and oxidizer. The flow is controlled by solenoid valves, driven from the computer. Generally, accelerometers and gyros monitor the jet firing to see if the desired effect has been achieved. If not, a second firing can be used to get the results closer to the desired.

Housekeeping tasks

The spacecraft onboard computer monitors onboard systems, and takes corrective action as needed. It monitors a variety of subsystems.

Generally, there is a dedicated unit, sometimes referred to as the Command & Data Handler (C&DH) with interfaces with the spacecraft transmitters and receivers, the onboard data system, and the flight computer. The C&DH, itself a computer, is in charge of uplinked data (generally, commands), onboard data storage, and data transmission. The C&DH can send received commands directly to various spacecraft components, or can hold them for later dissemination at a specified time. The C&DH has a direct connection with the science instrument(s) for that data stream. If the science instrument package has multiple sensors, there may be a separate science C&DH (SC&DH) that consolidates the sensed data, and hands it over to the C&DH for transmission to the ground. It is also common for the C&DH to hand over all commands related to science instruments to the IC&DH.

The C&DH will provide services to the IC&DH and the instruments in general, such as electrical power, pointing (although the instrument may be independent of the spacecraft pointing, and data services, such as onboard storage, and downlink.

Consumables inventory and management

The spacecraft computer calculates and maintains a table of consumables data, both value and usage rate. This includes available electrical power in the batteries, state-of-charge, the amount of thruster propellant remaining, and the status of any other renewable or consumable asset. This is periodically telemetered to the ground. Over the long term, we can do trending on this data, which can help us identify pending problems.

Thermal management

The spacecraft electronics needs to be kept within a certain temperature for proper operation. Generally, the only heat source is the Sun, and the only heat sink is deep space. There are options as to how the spacecraft can be oriented. In close orbit to a planet, the planet may also represent a heat source. Automatic thermal louvers can be used to regulate the spacecraft internal temperature, if they are pointed to deep space. The flight computer's job is to keep the science instrument or communications antennae pointed in the right direction. This might be overridden in case the spacecraft is getting too hot or too cold.

Electrical Power/energy management

The Flight Computer needs to know the state-of-charge (SOC) of the batteries at all times, and whether current is flowing into or out of the batteries. It the SOC is getting too low, some operations must be suspended, so the solar panels or spacecraft itself can be re-oriented to maximize charging. In some cases, redundant equipment may be turned off, according to a predetermined load-shedding algorithm. If the spacecraft batteries are fully discharged, it is generally the end of the mission, because pointing to the Sun cannot be achieved, except by lucky accident. Don't bet on it.

Antenna Pointing

The spacecraft communications antennae must be pointed to the large antennae on the ground (Earth) or to a communications relay satellite in a higher orbit (for Earth or Mars). The antenna can usually be steered in two axis, independently of the spacecraft body. This can be accomplished in the Main flight computer, or be a task for the C&DH unit.

Radiation Damage Mitigation

Homeostasis refers to a system that monitors, corrects, and controls its own state. Our bodies do that with our blood pressure, temperature, blood suger level, and many other parameters.

We can have the spacecraft computer monitor its own performance, or have two identical systems monitor each other. Each approach has problems. We can also choose to "triplicate" the hardware, and use external logic to see if results differ. The idea is, two outweigh one, because the probability of a double error is less than that of a single error.

In at least one case I know of, the backup computer erroneously thought the primary made a mistake, and took over control. It was wrong, and caused a system failure.

To counter the effects of "bit flips" and other effects of radiation, the memory can be designed with error detection and correction (EDAC). Generally, this means a longer, encoded word that can detect and correct M errors. There is a trade-off with price. With EDAC memory, there is a low priority background task running on the cpu that continuously reading and writing back to memory. This process, called "memory scrubbing" will catch and correct errors.

Self-test software can be included, usually running as a background task. This might also send a "heart-beat" signal to another processor or logic. At the hardware level, particularly if we are using configurable logic, we can include built-in self test (BIST).

Safe Hold mode

As a last resort, the spacecraft has a safe-hold or survival mode that operates without computer intervention. This usually seeks to orient the spacecraft with its solar panels to the Sun to maximize power, turn off all non-essential systems, and call for help. This can be implemented in a dedicated unit.

Flight software

All spacecraft are run by on-board computers that implement the tasks described above. A set of common spacecraft fuctions has been implements in an open-source library from NASA's Goddard Space Flight Center, Flight Software Branch. This is called the Core Flight Software. It runs under the Core Flight Executive (CFE). The executive is a set of mission independent reusable software services and an operating environment. Within this architecture, various mission-specific applications can be hosted. The cFE focuses on the commonality of flight software. The Core Flight System (CFS) supplies libraries and applications. Decades of flight software legacy went into the concept of the cFE. Various modules provide functions such as command and telemetry service, scheduling, limit checking, file delivery in CCSDS format, etc.

Flight CPU's

On regular space missions, the use of a Radiation-hardened flight computer is essential. This can be orders of magnitude more expensive that the terrestrial equivalent, and the state of the art lags commercial practice by at least one hardware generation. The processor of choice to go to the outer planets is the RAD-750's, essentially a rad-hard PowerPC architecture as found in Apple Compuers.

For Cubesat missions, a Raspberry PI arehitecture is favored. Besides the standard 32-bit integer and floating point data structures, the cpu may contain specialized processing units.

Orbit and Ephemeris

This section serves as an introduction to satellite orbits, essentially the path the satellite follows. The path is set by the launch vehicle, but is affected by several factors. First, the Earth has a non-uniform gravity field. As you fly over the Himalaya's the gravity force is greater than when you fly over the ocean. The moon affects the orbit – it produces the tides by pulling up on the oceans, after all.

We talk about zero gravity in space. That's not quite right. A satellite in orbit is in the Earth's gravity field, that of the moon and the Sun, and actually every other object in space.

Basically, there is gravity in space, but the satellites are in free-fall. They appear to be weightless, but their "mass" remains the same as it always did. Your weight depends on your mass, the local gravity field, and how far away you are from the much larger mass (the Earth). If you travel about the solar system visiting planets and moons, your mass remains the same, but you weight varies with the size of the planet or moon. Mass is an intrinsic feature of matter. Weight is related to the gravitation field we live in. Isaac

Newton gave us this insight in the 1700's.

Celestial Mechanics

(does a celestial mechanic carry a space wrench?)

ok, this is a complicated topic if you haven't had a lot of physics and math, so I'll try to go slowly.

How do we describe the position and velocity of something in orbit? We start with F=MA (force = mass times acceleration). Force and acceleration will be vector quantities in three dimensions. Vector quantities have both a numeric value, and a direction. The mass is an intrinsic property of the satellite.

Ok, we have three vector quantities. To get from the acceleration to the position, we need second order differential equations. (Don't panic). What this means is, we have (3 x 2) or six parameters to define the position and velocity of the spacecraft in space. These parameters will be with respect to a selected frame of reference. We can put a frame of reference wherever we want. Earth-centered, Sun-centered, our backyard. Its arbitrary, but we choose a convenient spot, and measure distances and such in that frame.

Now, we are indebted to Johannes Kepler for his work on orbits in the 1700's. He was working on the orbits of the planets around the Sun. His pick of the six parameters to describe a body's position in orbit have become known as the "Keplerian" elements. There are lots of different ways to do it, but this one is the most common. Keep in mind, these six elements are only valid at a particular time, called the Epoch, and is continuously changing.

This is the solution to what is called the two body problem. Kepler assumed a large mass (the Sun) with much smaller masses (planets) orbiting it. You can also assume a body such as the Earth with small bodies like satellites orbiting it. Our satellite will also be affected by the moon, the Sun, and every other chunk of rock in our solar system. We can account for this with "the method of perturbations," meaning we calculate the ideal orbit, and figure the small changes due to other objects.

We're not going any farther in Orbital Mechanics. You can take the multi-Semester course, and there are a lot of good software tools available. You might even come up with the solution to the multi-body problem.

Orbital Decay

The atmosphere doesn't really stop abruptly at some point above the planet, it just gets thinner. So, there is always some drag involved from solar panels, or antennas. This is small, but it builds up over time. Large mission like the ISS adjust their orbits periodically, by firing boosters to take the assembly back to the correct altitude. Left uncorrected, the satellite will eventually re-enter the denser points of the atmosphere and burn, due to frictional heating. Actually, we plan for that to happen, and older satellites are deliberately de-orbited so they don't interfere with ongoing missions. Now, every spacecraft mission has to have an end-of-life plan. Even Cubesats. There is a product called "terminator tape" for Cubesats, that is deployed at the end of the mission. It creates additional drag, and causes more rapid orbital decay.

Talking and Listening to a Satellite.

Communicating with and operating a spacecraft in orbit or on another planet is challenging, but is an extension of operating any remote system. We have communications and bandwidth issues, speed-of-light communication limitations, and complexity. Remote debugging is a always a challenge. In most cases, we use an RF link, but laser communications is possible.

The satellite control center is part of what is termed the Ground Segment, which also includes the communication uplink and downlink. The control center generates uplink data (commands) to the spacecraft, and receives, processes, and archives downlink (telemetry) data. The spacecraft is usually referred to as the space segment. The spacecraft usually consists of a "bus", the engineering section, and the payload, either a science instrument package or a communications package. Satellite busses can be "off-the-shelf," leading to economies of scale. The new popular form factor is the Cubesat. The buss consists of the structure, thermal control, electrical power, attitude determination and control, and telemetry and command. It also interfaces to the launch vehicle. The bus hosts he science insturment package, providing it with services such as power and telemetry and comand.

A satellite control center has a wide variety of tasks. It provides services to the mission 24x7x365. These services include the reception, archiving, limit-checking, and conversion of the received telemetry data. Today, received telemetry is archived in raw form, and saved in engineering units in a database. Use of a standard commercial database simplifies operations and controls costs. The Control Center disseminates selected data to users, either located in a control room, or via the web. The Control Center will use automation of operations where possible. The software does limit checking of incoming data, and issues alerts if limits are exceeded. The control center also is responsible for commanding the spacecraft.

It is easy to implement a satellite control center – on a laptop. The *COSMOS* product from Ball Aerospace in Boulder Colorado, is free and open source. For simple missions like a Cubesat, it is adequate. How do you get your satellite data, and how do you send commands to the satellite? Well, you can implement your own ground station. An open source ground station is available from SatNogs. You build the antenna and receiver. Easy as can be.

NASA's unmanned near-earth missions are the responsibility of the Goddard Space Flight Center in Greenbelt, Maryland. Initially, control centers were as different and unique as the early spacecraft themselves. As more satellite series came along (TDRSS, NOAA weather satellites) there was a corresponding similarity in control centers.

For U. S. missions beyond the near-Earth, the responsibility falls to California Institute of Technology's Jet Propulsion Laboratory, under contract to NASA. The control center's for the various missions resemble those used at Goddard Space Flight Center for near Earth missions. They are purpose-built, since each mission is highly unique. Some JPL missions operate for decades or multiple decades, so a technology refresh is sometimes required. Their Mission control rooms are called OCC, Operations Control Centers, and the Space Flight Operations Facility, SFOF. Manned flight is handled at the Johnson Space Center in Texas.

Satellites are tracked by radar. NORAD, the North American Aerospace Command, based in Colorado, tracks all detectable orbital entities, from large satellites to space junk, zombie-sats, and the larger pieces of debris, as well as near-Earth asteroids.

NORAD puts all this up on a website for your convenience, in a standard format called the "two-line element" (TLE). This contains the Keplerian orbital elements, the set of data describing the orbit of anything around the Earth, for a given point in time (epoch). You have a world-wide tracking infrastructure available, with the data downloadable from a website.

Living and working in Space – the ISS

The International Space Station is a research facility in Earth orbit.It is permanently crewed, with crew change every 6 months ot a year. Some tourists have paid (big money) to visit, but that porgram is suspended at the moment due to the grounding of the Shuttle fleet. Transportation is the issue, and the Russian Soyuz capsule is the only option at the moment.

To this writing, Space Adventures is the only private company that has sent paying customers to space. They work with NASA, the Russian Space Agency, and other private space companies. It was founded in 1998, and has offices in Tyson's Corner, Virginia. The company is made of of entrepreneurs from the adventure travel, entertainment, and aerospace idustries.

How do I go to the bathroom?

On the Space Shuttle, it was just inside the hatch, to the right. On the Space Station, it is in the Russian Zvezda module. A zero-g toilet takes some getting used to. There is also a shower.

What time is it?

Since the Space Station goes around the Earth in about 90 minutes, your "day" is 45 minutes, give or take, and so is your "night." You can try to keep on a back-home schedule. Most orbital missions use GMT as a standard time.

How do I sleep?

Most astronauts on the Space Station prefer to Velcro themselves to a wall, in a sleeping bag.

Can I go outside?

Yeah......with a bit more training on the use of a spacesuit, and a lot more money . You will need a tether, to hold yourself to something , else you will float off, and become a satellite yourself.

Should we be spending money in Space?

Trick question? There is nowhere to spend money in space. The International Space Station doesn't have an ATM. We spend money on Earth for space related projects. Doing this gives us a huge return. Think about weather forecasting, the GPS navigation system, and Dish Network.

Getting fancy, multiple units.

We are not limited to one satellite. We can have groups of satellites complementing each other. There are several ways to accomplish this.

Constellations

Constellations are groups of satellites operating together to observe a single target A constellation allows you to do simultaneous observations of one target from multiple locations, or multiple targets simultaneously. The elements of a constellation can be homogeneous or not.

NASA says, " A Constellation is a space mission that, beginning with its inception, is composed of two or more spacecraft that are placed into specific orbit(s) for the purpose of serving a common objective (e.g., Iridium)." (The Iridium constellation is for communications).

Other examples include the Galileo Constellation of European navigation satellites, the US GPS navigation satellites, and NASA's Tracking & Data Relay Satellites, at Geosynchronous orbit. Internet via satellite link is provided to the scientific community in Antarcticia by a constellation of polar orbiting comm-sats.

Clusters

A cluster means a group of loosely coupled elements, working together on the same problem. If the cluster were more tightly coupled, and self-directed, we'd have a swarm. As it is, we have a group of individuals that could be considered a single entity. Generally, members of a compute cluster have the same hardware and software configuration. One issue in clustering is the degree of coupling between elements. No coupling means we have a mob. A lot of coupling and we might have a swarm. In a cluster, management of the cluster itself can be centralized in a control unit, or can be distributed across the cluster.

Trains

Trains of satellites refer to multiple units that are spaced along the same orbital track. This allows for simultaneous imaging areas, as well as continuous observation of selected areas. It is a co-ordinated group of observation satellites. NASA uses this approach successfully for Earth Science and Weather satellites in polar, sun-synchronous orbits. A train of 6 weather satellites passes over the same spot at the same time every day. The satellites are all different, but provide useful information on atmospheric and ground conditions. Another term for a simple group of spacecraft in the same orbit is *string-of-pearls*.

A single orbit with four sun-synchronous Disaster Monitoring Constellation (DMC) satellites was implemented in 2002-2003. A second Constellation was implemented with 4 additional satellites by 2011, with participation by the U.K., Algeria, Nigeria, Turkey, and China.

Exploring our solar system – the complexity of the Problem.

Let's look at the number of objects in our solar system that we would like to know more about. We will also list the one-way light times for the various objects. This tell us how long a radio signal takes to traverse that distance.

Trojan Asteroids are in the same orbit as a planet, in front and behind.

Earth – 1 moon, 1 Trojan. Earth's moon is about ½ light second away.

Near-Earth Objects (NEO's) – more than 15,000.

Technically, an NEO is a solar system object whose closest approach to the Sun is 1.3 AU, and that comes in close proximity to the Earth There are 14,000 known asteroids in this category, 100 comets, solar orbiting spacecraft, and meteoroids. All these have the potential of striking the Earth. They are closely tracked from the ground, by NASA's Planetary Defense Coordination Office. A joint US/EU project called Spaceguard is tracking NEO's larger than 30 meters. Three NEO's have been visited by spacecraft.

Mercury - no moons. The planet is in tidal lock with the Sun. It wiggles a bit, creating a twilight zone where the temperature extremes are not as bad.

Venus – no moons. Has Trojans.

Asteroid belt – Ceres the dwarf planet, and 750,000 rocks larger than 1 km. The nearest point of the asteroid belt to Earth is about 1.2 AU.

Mars, its 2 moons and 7 Trojans. On-way light time varies between 3 and 29 minutes.

Jupiter and its 67 moons, 6,000 Trojans, 33-53 minutes one-way light time.

Saturn and its 62 moons, 1.4 hours one-way light time.

Uranus and it's 27 moons, 2.7 hours one-way light time. 1 known Trojan.

Neptune and its 14 moons, 4.3 hours one-way light time, 18 Trojans.

Pluto and its moons Charon, Nix, Hydra, Styx and Kerberos, 4.6 hours one-way light time.

Comets – 5, 253 known.

Centaurs – icy minor planets between Jupiter and Neptune – there are 3 known.

The Kuiper Belt extends from the orbit of Neptune out approximately 50 AU. There are five known dwarf planets, the former planet Pluto, and Ceres, Haumea. Makemake, and Eris. Over 100,000 objects are speculated to exist. Neptune has a major influence over the Kuiper belt objects. Not much is known about the belt and its Trans-Neptunian Objects (TNO's), since astronomers so far have had to rely on ground based observation. The New Horizons mission is currently proceeding out through the Kuiper belt, and is reporting back what it sees.

Beyond the Kuiper belt is the Scattered disc, extending beyond 100 AU. This is a sparsely populated region of the solar system. This is no knowledge of how many scattered disk objects (SDO's) exist. The closest is at around 30 AU. The belt extends above and below the ecliptic plane, in a torus configuration. The ecliptic plane is the apparent path of the Sun around the Earth. Most of the planets in the solar system orbit in ths plane.

Beyond this is the Oort Cloud, extending out 5,000 – 100,000 AU, There is a disk-shaped inner cloud, and a spherical outer cloud. And you thought space was empty.

We need to get a lot of spacecraft working on these explorations. U.S. Spacecraft have visited all of the planets in the Solar System (including the demoted Pluto).

Exploring the Sun

We get all our energy from this huge thermo-nuclear reactor some 8 light-minutes away. As you get closer to the Sun, its gets hotter, and there is more radiation in terms of energetic particles.

NASA sent a series of probes to observe the sun in 1958 and for ten years after. These were the Pioneer 5-9 spacecraft. They didn't get any closer, but provided a different point of view. They got good data on the solar wind and the sun's magnetic field. Pioneer 9 sent back good data for 15 years. The Helios spacecraft in the 1970's were joint U.S.-German missions that used an orbit that got within the orbit of Mercury.

The 1980 Solar Maximum Mission observed the Sun in the spectrum of gamma rays, X-rays, and Ultraviolet. SMM had a failure in its electronics months after launch, but was repaired by a subsequent Shuttle mission.

A Japanese Mission to study the Sun was Yohkoh, or Sunbeam, in 1991. It imaged solar flares in the x-ray spectrum.

Other US missions included SOHO, the Solar and Heliospheric Observatory, and the Solar dynamics Observatory. SOHO was located at the Lagrangian point between the Sun and the Earth, which is a null point in the gravity field. It sees the Sun constantly in many selected wavelengths.

The Solar Terrestrial Relations Observatory (Stereo) is a dual spacecraft mission to the Sun, launched in 2006. One is ahead of the Earth in orbit, the other behind. This provides three points of view of solar phenomena.

The Ulysses spacecraft, discussed later in the section on Jupiter, left the plane of the ecliptic (thanks to the Jupiter swing-by) and observed the Sun's high latitudes. One of its discoveries was that large magnetic waves emitted from the Sun scattered galactic cosmic rays.

The Genesis mission was designed to capture and return solar material. It achieved its goal, but was damaged in a crash landing when it returned to Earth in 2004.

Two areas that have never been imaged are the Sun's polar regions. This is because of the very large energy expenditure required to get out of the plane of the ecliptic. The process of achieving an out-of-ecliptic plane and into a path that gets us to the solar poles continues as a research topic, with no known solutions.

A proposed mission, the Solar Polar Constellation, would be dedicated to high inclination solar orbit.

Exploring Mercury

The U. S. Messenger mission to Mercury, the closest planet to the Sun, was launched in 2004. It is currently orbiting the hottest planet. No landing on Mercury has been attempted, although it would be feasible in the *twilight zone* between the extremely hot solar facing side, and the much colder space facing side. Mercury is in tidal lock with the sun, with one side always facing it. It wobbles a bit, creating a "twilight zone" that is much less extreme. It has no known moons, or Trojans. Being so close to the Sun, it is difficult to observe the planet and its immediate vicinity.

Exploring Venus

The Soviet space program sent a series of probes to Venus. Early efforts were either crushed in the dense atmosphere, or suffered thermal damage. The Venera-7 mission had a goal of surface sample return. It struck the surface harder than planned, but returned temperature data for about 20 minutes. The Venera-8 probe returned data for some 50 minutes. Venera-13 and -14 returned color photos of the surface. Further Soviet and US efforts involved observation from Venus orbit. The Venus environment has proven extremely hostile. It seems our sister world, next towards the Sun from us, is in a environmental runaway condition. Heavy greenhouse clouds trap the solar energy, and cause massive global warming on a planetary scale. The surface temperature is high enough to melt some metals. This is very hard on computers, and electronics in general.

Venus' atmosphere is 96% carbon dioxide at a surface pressure of nearly 100 times Earth's, a greenhouse gone wild. It has no moons. Venus is roughly Earth-sized, but something went terribly wrong. It also has clouds of sulphuric acid, that landers have to get through. There is no magnetic field, but there is active volcanism.

Venus express, an ESA mission, is in Venus Polar orbit. It found a massive double atmospheric vortex (storm) at the south pole. Venus Express operated from 2005-2014. Venus has no moons, but does have Trojans. The Japanese Venus Climate Orbiter "Akatsuki

" was launched in 2010, but failed to achieve Venus orbit. It orbited the Sun for 5 years, and was finally put into Venus orbit in 2015.

Near Earth Objects

Technically, an NEO is a solar system object whose closest approach to the Sun is 1.3 AU, and that comes in close proximity to the Earth There are 14,000 known asteroids in this category, 100 comets, solar orbiting spacecraft, and meteoroids. All these have the potential of striking the Earth. They are closely tracked from the ground, by NASA's Planetary Defense Coordination Office. A joint US/EU project called Spaceguard is tracking NEO's larger than 30 meters. Three NEO's have been visited by spacecraft.

Cubesats, with solar sails, are an ideal approach to explore objects in our home vicinity, because of the danger they may express, but also for possible exploitation by mining.

Exploring the Asteroid Belt, dwarf planets, and Centaurs

Asteroids have been imaged by the New Horizons spacecraft, on its way to Pluto, and by the Cassini spacecraft. The Pioneer-10 spacecraft was sent to study the far reaches of the solar system It passed through the Asteroid belt on its way to Jupiter and Saturn.

A driver in the space environment is the exploration of the asteroids, numbering in the thousands. Although there are fewer than 10 planets, and less than 200 moons, there are millions of asteroids, mostly in the inner solar system. The main asteroid belt is between Mars and Jupiter. Each may be unique, and some may provide needed raw materials for Earth's use. There are three main classifications: carbon-rich, stony, and metallic.

The physical composition of asteroids is varied and poorly understood. Ceres appears to be composed of a rocky core covered by an icy mantle, whereas Vesta may have a nickel-iron core. Hygiea appears to have a uniformly primitive composition of carbonaceous chondrite. Many of the smaller asteroids are piles of rubble held together loosely by gravity. Some have moons themselves, or are co-orbiting binary asteroids. The bottom line is, asteroids are diverse.

It has been suggested that asteroids might be used as a source of materials that may be rare or exhausted on earth (asteroid mining) for materials for constructing space habitats or as refuelling stations for missions. Materials that are heavy and expensive to launch from earth may someday be mined from asteroids and used for space manufacturing. Valuable materials such as platinum may be returned to Earth for a profit.

There are only 8 ½ planets, but there are thousands of asteroids, and it seems there may be as many types. This means that exploring the known asteroids is a daunting challenge. On the other hand, the asteroids can be a significant source of raw materials for Earth. A conventional survey and exploration approach would take too long. What is needed instead is a multitude of autonomous and flexible nano-spacecraft. The architectural model is a swarm (social insect model) distributed intelligence. The platform of low cost, low power, low weight could involve Cubesats with solar sails. The asteroid belt contains Ceres, the Dwarf planet, and some 750,000 rocks larger than one kilometre in diameter.

The physical composition of asteroids is varied and poorly understood. Ceres appears to be composed of a rocky core covered by an icy mantle, whereas Vesta may have a nickel-iron core. Hygiea appears to have a uniformly primitive composition of carbonaceous chondrite. Many of the smaller asteroids are piles of rubble held together loosely by gravity. Some have moons themselves, or are a co-orbiting binary pair. The bottom line is, asteroids are numerous and diverse.

The asteroids are not uniformly distributed. In the asteroid belt, the Kirkwood gaps are relatively empty spots. This is caused by orbital resonance of the asteroids with Jupiter. Orbiting irregular shaped bodies is challenging, due to the irregular gravity field. This makes station keeping and attitude control a problem.

A Centaur is a type of dwarf planet, not quite making the cut to "real planet." there are 44,000 known examples with diameters greater than 1 km. They have unstable orbits that intersect those of the gas giants. They are somewhat like asteroids, and somewhat like comets. The largest known, Chariklo, has a ring system. These have not been photographed from a close position.

Exploring Comets

There are some 5,253 known comets. The Deep Impact mission returned images of the surface of comet Borrelly in 2001. That surface was hot (26-70C), dry, and dark. In July of 2005, the same mission sent a probe into Comet Tempel 1. It created a crater, allowing imaging of subsurface material. Water ice was seen. Comet Borrely has a coma (tail), which proved to be vaporized subsurface water ice. Deep Impact went on to complete a flyby of Comet Hartley-2 in 2010.

The 1999 Stardust mission retrieved sample material from the tail of Comet Wild 2 and returned it to Earth in 2006. It released a lander, Philae, which successfully touched down on the comet's surface in 2014.

Pioneer Venus observed Comet Halley while in transit. This was during a period when the comet was not visible from Earth, because of its proximity to the Sun. The Venus probe monitored the loss of water from the comet as it swung around close to the Sun.

Exploring Mars

Mars, and its two tiny moons and seven Trojans has got some infrastructure in place – a communications relay satellite and a weather satellite. There are several Rovers and landers on the surface.

The Viking program was a pair of spacecraft sent to Mars in 1975. Each spacecraft consisted of an orbiter, and a lander. A major target is a Mars sample return mission.

The Mars Pathfinder mission landed on Mars on July 4, 1997. It carried a Rover named Sojourner, which was a 6-wheeled design, with a solar panel for power, but the batteries were not rechargeable. The rest of the lander served as a base station. Communication with the rover was lost in September. It communicated with Earth via the base station using a 9600 baud UHF radio modem. The communication loss leading to end of mission was in the base station communication, while the Rover itself remained functional. The Rover had three cameras, and an x-ray spectrometer.

The MER (Mars Exploration Rovers *Spirit & Opportunity*) are six-wheeled, 400 pound solar-powered robots, launched in 2003 as part of NASA's ongoing Mars Exploration Program. *Opportunity* (MER-B) landed successfully at Meridiani Planum on Mars on January 25, 2004, three weeks after its twin *Spirit* (MER-A) had landed on the other side of the planet. Both used parachutes, a retro-rocket, and a large airbag to land successfully, after transitioning the thin atmosphere of Mars.

The Spirit unit became stuck in 2009, and engineers were unable to free it after 9 months of trying. It was re-tasked as a stationary sensor platform. Contact was lost in 2010.

This is an ongoing mission. It was originally planned for 90 days, but the *Opportunity* Rover is still collecting useful data regarding potential life on our sister planet some 11 years later as of this writing. It has traveled over 35 kilometers on the Martian surface.

The Mars Science Laboratory landed successfully on the Martian surface on August 6, 2012. It had been launched on November 26, 2011. It's location on Mars is the Gale crater. It is designed to operate for two Martian years (sols). The mission is to determine if

Mars could have supported life in the past, which is linked to the presence of liquid water.

The Rover vehicle *Curiosity* weights just about 1 ton (2,000 lbs, 900 kg.) and is 10 feet(3 meters) long. It has autonomous navigation, and is expected to cover about 20 km over the life of the mission. The platform uses six wheels

Communication with Earth uses a direct X-band link, and a UHF link to a relay spacecraft in Mars orbit. At landing, the one-way communications time to Earth was 13 minutes, 46 seconds. This varies considerably, with the relative positions of Earth and Mars in their orbits around the Sun.

The science payload includes a series of cameras, including one on a robotic arm, a laser-induced laser spectroscopy instrument, an X-ray spectrometer, and x-ray diffraction/fluorescence instrument, a mass spectrometer, a gas chromatograph, and a laser spectrometer. In addition, the rover hosts a weather station, and radiation detectors. There is cooperation between in-space assets and ground rovers in sighting dust storms by the meteorological satellite in Mars orbit.

NASA's Maven (Mars Atmosphere and Volatile EvolutioN Mission) mission to Mars is an orbiter, to study the Martian atmosphere It was launched in November 2013, and reached Mars in September of 2014. It is still operating as of this writing.

Exploring the Gas Giants
The Gas giants are the planets Jupiter, Saturn, Uranus, and Neptune. These are the responsibility of the Jet Propulsion Laboratory. These each have extensive ring and moon systems that have been imaged, but not yet explored.

Pioneer 10 was the first mission to Jupiter, followed by Pioneer-11 in 1973, and, as of this writing, there have been 8 total. Jupiter has a very high trapped radiation environment. They are mostly all different, and some are thought to be capable of hosting life, as we know it. The moon Io has volcanic activity, and Europa has water ice on the surface. Europa is considered "one of the most promising extraterrestrial habitable environments in our solar system" according to the most recent Planetary Society's Decadal Survey. A proposed mission, ExCSITE, would provide characterization of the surface properties. One option is multiple Cubesats.

NASA is looking at the Explorer CubeSAt for Student Involvement in Travels to Europa (ExCSITE) Mission. This involves multiple Cubesat imagers and impactors.

The Voyager missions were originally terms the "Grand Tour" and were to have visited Mars, Jupiter, and Saturn, with possibly some of the outer planets as well. The mission was called MJS-77. Budget constraints caused the mission to refocus on Jupiter and Saturn alone. The author worked on this mission.

Jupiter

Jupiter has 67 known moons, and perhaps 1 million Trojans of 1 kilometer or larger. These tend to congregate at L4 and L5. The largest has a diameter of several hundred kilometers. The one way light time for Jupiter is 33-53 minutes,

Cassini observed the planet from close-up in the year 2000, and studied the atmosphere. Galileo entered Jupiter orbit in 1995, and returned data on the planet and the four Galilean moons until 2003. Three of the moons have thin atmospheres, and may have liquid water. The moon Ganymede has a magnetic field. Galileo was in the right place at the right time to see the comet Showmaker-Levy-9 enter the Jovian atmosphere, and launched an atmospheric probe.

The Juno mission to Jupiter has just arrived after 5 years of travel, and was getting settled in to begin its observations. This project was launched in August of 2011, and arrived at Jupiter in July 2016. It was placed in Jupiter elliptical polar orbit for 5 years, and will de-orbit into Jupiter in February 2018. This is to ensure burn-up of the spacecraft to avoid any biological contamination of Jupiter or its moons. It is scheduled to may 37 orbits. The orbit was chosen to minimize contact with Jupiter's intense trapped radiation belts. It's sensitive electronics are housed in "the Juno Radiation vault," with 1cm titanium walls. It will have available to it some 420 watts of power, from the solar arrays.

The spacecraft weighs over 1,500 kg. It uses 3 solar panels of 2.7 x 8.9 meters long These will be exposed to about 4% of the sunlight at Earth. It left Florida on an Atlas-V vehicle. The perijove, or closest distance to the planet was planned to be 4,200 km. The highest altitude at apojove is 8.1 million kilometers.

It includes infrared and microwave instruments to measure the thermal radiation from Jupiter's atmosphere, being particularly interested in convection currents. It's data will be used to measure the water in Jupiter's atmosphere, and measure atmospheric temperature and composition, and track cloud motions. The mission will also map Jupiter's magnetic and gravity fields. It is expected to probe the magnetosphere in the polar regions and observe the auroras.

Communications uses X-band to support 50 Mbps of data. The spacecraft will be constrained to 40 Mbytes of camera data per 11-day orbit period.

Juno uses a bi-propellant propulsion system (for insertion manoeuvrers) and a monopropellant system for attitude control.)

Saturn

Saturn and it's 62 known moons has a one-way light time around 1.4 hours. Saturn has been visited by spacecraft four times. The first was a flyby by Pioneer-10 in 1979. This showed the temperature of the planet was 250 degrees K. Voyager-1 visited in 1980. It conducted a close flyby of the moon Titan to study its atmosphere. It is, unfortunately, opaque in visible light. We do know it rains methane. Voyager-2 swung by a year later, and data showed changes in the rings since its sister mission visited the year before. Temperature and pressure profiles of the atmosphere were gathered. Saturn's temperature was measured at 70 degrees above absolute zero at the top of the clouds, and -130 c near the surface. The flybys discovered additional moons, and small gaps in the rings.

Cassini was the fourth spacecraft to study Saturn, which has rings, although smaller than Jupiter. The rings were confirmed by the Voyager spacecraft in the 1980's. Cassini entered into Saturnian orbit, and is still returning data. The one-way communications time varies form 68-84 minutes. It has also collected data on the Saturnian moons Titan, Enceladus, Mimas, Tethys, Dione, Rhea, Iapetus, and Helene. Things are strange in the Saturnian system. Cassini observed a hurricane in 2006 on the planet's south pole. It appears to be stationary, 5,000 miles (8,300 km) across, 40 miles (67 km) high, with winds of 350 mph (560 kph). The large moon Titan has lakes of a liquid hydrocarbon, with possible seas of methane and ethane. Cassini launched a probe *Huygens* to Titan, and it landed on solid ground below the atmosphere. The Cassini mission was responsible for the discovery of seven new moons of Saturn.

Cassini observed a massive storm on Saturn, the great white spot, that recurs every 30 years. The storm, larger than the red one on Jupiter, exhibited a discharge that spiked the temperature 150 degrees. At the same time, Earth observations showed a large increase in atmospheric ethylene gas. It also discovered large lakes or seas of hydrocarbons near the planet's north pole.

Cassini discovered a possible atmosphere on the moon Enceladus, with ionized water vapor, and ice geysers. Many of the Saturnian moons are in tidal lock with their mother planet. Being so close to its giant neighbor Jupiter affects the Saturnian system.

Uranus

Uranus has 27 known moons, a 13-ring system, and a one way light time of 2.7 hours from Earth. It has one known Trojan. Uranus was imaged in a flyby by the Voyager-2 spacecraft in 1986. It also captured some images of the Uranian moon Umbriel. But, Uranus and Neptune are one of the great remaining unknowns in the solar system, since neither have been explored in detail, by a dedicated mission. There is a desire to put an explorer spacecraft in orbit, and use that as a platform to launch probes into the atmosphere.

Uranus and Neptune are sometimes referred to as ice giants, since their atmospheres are known to contain water, ammonia, and methane ice. Uranus has a magnetic field. Interestingly, Uranus' spin axis is tilted into the plane of its orbit around the Sun. Seasonal changes and weather have been observed. The Voyager-2 mission imaged Uranus on its way from Jupiter, and out of the solar system. Atmospheric wind speeds are know to approach 900 kilometers per hour. It's orbit period is 84 Earth years. It receives about $1/400^{th}$ of the light that the Earth does from the Sun, so solar power is probably not a viable choice.

Because of the strange orientation of the planet's rotation axis, during the solstice, one side of the planet faces the Sun continuously, and the other faces deep space. Each pole gets 42 years of direct (though weak) sunlight, and 42 years of darkness. In spite of this, the equator is the hottest region. At this writing, the planet is in its autumnal equinox.

Uranus has a strange predominately water-ammonia ocean, which is electrically conductive. A major targeted mission is the Uranus orbiter and Probe. Mission analysis comes up with a 12-13 year cruise from Earth to Uranus.

Neptune

Neptune has 14 known moons, and 18 known Trojans. It's one-way light time is around 4.3 hours. Neptune has also been visited by Voyager-2 in 1989. It discovered six new moons. That is the extent of close-up observations of the planet. Neptune has rings, like Jupiter and Saturn, and a great dark spot. It's moon Triton has geysers and polar caps. Triton has an interesting retrograde orbit – it goes in a different direction than the other moons. Triton's surface is mostly frozen nitrogen, and is geologically active. It is speculated that Triton has a subterranean ocean. The moon Ptoteus is an ellipsoid, not a sphere.

Exploring Pluto, and beyond

Pluto was downgraded from a planet to a Kuiper Belt object. The New Horizons mission to Pluto and the Kuiper Belt began in January of 2006, and reached the vicinity of Pluto in July 2015. It conducted a 6-month survey of Pluto, and went out farther into the Kuiper belt, on an 3 year extended mission, which is ongoing at this writing.

To conserve heat and mass, New Horizon's spacecraft and instrument electronics are housed together in IEMs (Integrated Electronics Modules). There are two redundant IEMs.

In March of 2007, the Command and Data Handling computer experienced an un-correctable memory error and rebooted itself, causing the spacecraft to go into safe mode. The craft fully recovered within two days, with some data loss on Jupiter's magneto-tail. The one-way light time back to Earth is 4.6 hours.

In 2015, the Pluto flyby occurred, and data began to flow back to Earth. It took a year for all the imaging data to be transmitted, due to distances and transmit power involved.

Pluto had one known moon, Charon, before New Horizons Team members, using Hubble Space Telescope data, discovered Nix, Hydra, Styx, and Kerebos.

Kuiper Belt Objects (KBO)

The Kuiper Belt extends from the orbit of Neptune out approximately 50 AU. There are three known dwarf planets, the formal Planet Pluto and two others. Over 100,000 units are speculated to exist. Neptune has a major influence over the Kuiper belt objects. Not much is known about the belt and its objects, since astronomers have had to rely on ground based observation. The New Horizons mission is proceeding out through the Kuiper belt, and will tell us what it sees.

That's just the neighborhood. You want to talk about interstellar missions now? In space, particularly behind the Moon, when the Sun's light is blocked, is a good observation point.

The James Webb Space Telescope is the follow-on to the current Hubble Space Telescope. The project began in 1996, and it currently has a schedule launch in 2019. It will be located at the Earth-Moon L2 Lagranian point, which is behind the moon. It can observe in the red visible through near-infrared wavelengths. It will be operated from the Space Telescope Science Institute on the campus of Johns Hopkins University, in Baltimore, Maryland. The new telescope has 18 mirror segments, which are all adjustable, for a total of 25 square meters.There are secondary and tertiary mirrors. It has a large Sun shield, which will keep the mirror and science instruments at -370 degrees. The instruments include a near-infrared camera, a near-infrared spectrograph, and a mid-infrared instrument. Since very little infrared light penetrates the atmosphere, we will certainly be seeing the Universe "in a new light."

The spacecraft was built at the NASA-Goddard Space Flight Center in Greenbelt, MD. Early on, the project was known as the Next-generation s[pace telescope, but was named after the Apollo-era head of the agency. It is a joint project with the European Space Agency, and the Canadian Space Agency. Due to its planned location, repair is currently out of the question. The launch will be on an Ariane R rocket.

Manufacturing in Space

Manufacturing in Space is not that far away, and has actually been done on a small scale for many years. With permanent manufacturing facilities in space, near to lunar or asteroid resources, we will be able to fabricate facilities from local material, and extract rocket fuel. All of this can replace what we now need very large rockets up from Earth's "gravity well." We can build the next generation stations and spacecraft in situ, in orbit. There are some major advantages to this, as fewer parts need to be lifted up to orbit. Spin-off company's, providing logistics services, will be necessary. Space will be evolving as a frontier outpost. We have experience with those. But, space is a harsh environment, harsher than the Klondike during the gold rush. Yet, the gold rush happened.

Viewed in an economic sense, manufacturing "space stuff" in space makes sense. There is a large initial investment, but the reduced costs, particularly of things that are intended to stay in space, will more than balance this out.

Manufacturing of any kind, anywhere, involves the raw materials, a source of power, a source of labor (some may be robotic), and a transportation infrastructure. Following the 19th Century iron manufacturing model, it is better to manufacture near the source of the raw materials, and ship the finished product to the customer, than ship the raw materials. It is expected to be able to separate out usable amounts of iron, aluminum, silicon, and oxygen from lunar and asteroid material. In addition, water ice is know to exist at the lunar south pole. Extremely pure silicon wafers would be a valuable down-cargo. These could also be used in a subsequent process in-orbit to produce solar panels.

For space manufacturing, the power part is easy, sunlight 24x7. It is costly to bring materials up from the surface of Earth. The ideal situation is to use lunar or asteroid material. The customers may be on the Earth, or we may be building something that stays in orbit, or moves on to a different destination, such as Mars. It is expensive to boost skilled workers up from the surface, and keep them alive in orbit, but hopefully we can plan for tele-operation from Earth's surface, plus automated processes.

The rules change a bit in Space, and we will see yet another Industrial Revolution. Manufacturing in Space is by all definitions a Paradigm Shift. NASA's Marshall Space Flight Center in Huntsville, Alabama hosts the National Center for Advanced Manufacturing. Many commercial players are in this topic, including Bigelow, Orbital Technologies, Excalibur-Almuz, Space Island Group, and the Commercial Spaceflight Federation among others.

Space Tourism

There have been spacefarers from over 40 countries, taken along on shared missions by the craft of the major spacefaring nations, China, Russia, and the U. S. The International Space Station is truly an International effort. But these were all professional Astronauts or Cosmonnauts. That was their job. They got paid for it.

At this time, there have been seven "space tourists," who paid their own way, and five "spaceflgiht participants," who flew on the Shuttle, or to the ISS.

Can you fly to space now? The U.S. currently doesn't have a crewed transportation system. The Russians will charge you $76 million for a flight up on the Soyuz-M, if they have a seat available. You also receive training, and a couple of rides on their Vomit-Comet airplane, so you'll know what to expect in zero G.

The Space Tourism Industry is beginning. Like all new markets, it will evolve, become better and cheaper. It's expensive now, but a few have already done it. Options ranging from a quick trip above 100 km to earn the title "astronaut." to month-long vacations at a lunar resort, where you can fly, with wings, every day. Space based casinos and athletic venues are on the drawing board.

NASA is not going to do this. They are in the science and technology business, and are a government agency, A cadre of entrepreneurs, space geeks, and crafty businessmen have better, less expensive options in the works. Stay tuned. Keep in touch. This is going to get exciting.

Keep you eye on commercial outfits engaged in Space Tourism. It is going to be huge. With many commercial companies building launch vehicles, and particularily re-usable launch vehicles, the cost of these adventures will come down.

The allure of space will generate the interest, which will be tempered by the high cost of access. Think what you can do the Orbitel. You are weightless. You can see the stars most of the time. You are looking down n Earth, how cool is that? In a lunar resort, you can fly. Really. With wings, At 1/6 gravity, the spare-air tank will allow you to experience human-powered flapping wing flight. Actually, I think chickens would really enjoy the opportunity to finally fly. And the penguins!

Wrap-up

Hopefully, I have been able to clear up some key concepts in Space, and you are less consufed than before. If you can apply some of this to get some pre-K to 12's interested in Space, that would be a good thing. We need more knowledgable rocket-scientists and rocket engineers. The future is out there.

Bibliography

Aldrin, Buzz No Dream Is Too High, Life Lessons From a Man Who Walked on the Moon,National Geographic, 2016, ISBN-9781426216497.

Atkinson, Philip *Rocket men at Work and Play*, 2012, ASIN-B008DP24CQ.

Bektas, Metin *The Book Of Forces,* 2014, ASIN-B00P1Q0IR8.

Bora, C. J. *Introduction To Rocket & Space Craft Propulsion,* 2014, ASIN-B00P2TTR4O.

DK, *Space!, 2015,* Smithsonian, ISBN-1465438068.

Doody, Dave *Basics of Space Flight,* 2011, ISBN-0615476015.

Gilliland, Ben, *Rocket Science for the Rest of Us,* 2015, ISBN-1465433651.

Hill, Raymond *Spaceflight Theories: A beginners guide to rocket and space sciences,* 2014, ASIN-B00P2TTR4O.

Kastner, Bernice *Space Mathematics: Math Problems Based on Space Science,* 2013, ASIN-B00FC1N1LM.

Larson, Wiley J., Wertz, James R. *Space Mission Analysis and Design,* 3rd, ISBN-978-1881883104.

Motes, Andrew *Space Flight for Beginners,* 2015, ASIN-B019SOHMF4.

Rader, Andrew *Rocket Science,* 2017, ASIN-B078H81HZX.

Rogers, Lucy *It's ONLY Rocket Science: An Introduction in Plain English (Astronomers' Universe),* Springer, 2008, ISBN-038775377X.

Rowell, Rebecca *Building Rockets (Engineering Challenges),* 2017, ISBN-1635173205.

Schulte-Ladbeck, Dr. Regina E. *Basics of Spaceflight for Space Exploration, Space Commercialization, and Space Colonization,* 2016, ISBN-150252595X.

Stakem, Patrick H. *Cubesat Operations, How to fly a Cubesat,* 2017, PRRB Publishing, ISBN-152076717X.

Stakem, Patrick H. *Cubesat Engineering*, 2016, PRRB Publishing, ISBN-1520754019.

Stakem, Patrick H. *Cubesat Clusters and Swarms,* 2017, PRRB Publishing, ISBN-1520767544.

Stakem, Patrick H. *Interplanetary Cubesats*, 2017, PRRB Publishing, ISBN-1520896778.

Stakem, Patrick H. *Earth Rovers: for Exploration and Environmental Monitoring,* 2014, PRRB Publishing, ASIN B00MBKZCBE.

Stakem, Patrick H. *Spacecraft Control Center,* 2015, PRRB Publishing, ASIN-B01D1Y5LZ0.

Stakem, Patrick H. *Apollo's Computers*, 2014, PRRB Publishing, ASIN B00LDT217.

Stakem, Patrick H. *STEMsat, Using Cubesats in the pre-K to 12 STEM Curricula, A Resource Guide for Educators*, 2017, PRRB Publishing.

Stakem, Partick H. *Visiting the NASA Centers and Historic Rockets & Spacecraft,* 2017, PRRB Publishing, ASIN-B0757ZVB2G.

Swinerd, Graham *How Spacecraft Fly: Spaceflight Without Formulae,* Copernicus, 2009, ISBN 0387765719.

Taylor, Travis S. *Introduction to Rocket Science and Engineering*, CRC Press, 2009, ISBN-1420075284.

Taylor, Travis S. *A New American Space Plan,* 2012, ISBN-

1451638655.

Welti, C. Robert *Satellite Basics For Everyone: An Illustrated Guide to Satellites for Non-Technical and Technical People*, 2012, ISBN-147592593X.

Westerfield, Mike *Make: Rockets: Down-to-Earth Rocket Science*, 2014, ISBN-1457182920,

Resources

- NASA Systems Engineering Handbook, NASA SP-2007-6105. Avail: https://ntrs.**nasa**.gov/archive/**nasa**/casi.ntrs.**nasa**.gov/20080008301.pdf

- https://www.nasaspaceflight.com

- Encyclopedia Astronautica, http://www.astronautix.com/

- Vectors website - http://vc.airvectors.net/idx_sci.html

- https://history.nasa.gov/

- www.nasa.gov – NASA-knows

- https://spacemath.gsfc.nasa.gov/

- https://blogs.nasa.gov/

- Davis, Phillips."Basics of Space Flight".NASA., avail: https://solarsystem.nasa.gov/basics/

- Wikipedia, various

Glossary of terms

1U – one unit for a Cubesat, 10 x 10 x 10 cm.
3U – three units for a Cubesat
6u – 6 units in size, where 1u is defined by dimensions and weight.
802.11 – a radio frequency wireless data communications standard.
AACS – (JPL) Attitude and articulation control system.
ACE – attitude control electronics
Actuator – device which converts a control signal to a mechanical action.
Ada – a computer language.
A/D, ADC – analog to digital converter
AEB - Agência Espacial Brasileira
AFB – Air Force Base.
AGC – Automated guidance and control.
AIAA – American Institute of Aeronautics and Astronautics.
AIST – NASA GSFC Advanced Information System Technology.
ALU – arithmetic logic unit.
AmSat – Amateur Satellite. Favored by Ham Radio operators as communication relays.
Analog – concerned with continuous values.
ANSI – American National Standards Institute
Android – an operating system based on Gnu-Linux, popular for smart phones and tablet computers.
Antares – Space launch vehicle, compatible with Cubesats, by Orbital/ATK (U.S.)
AP – application programs.
API – application program interface; specification for software modules to communicate.
APL – Applied Physics Laboratory, of the Johns Hopkins University.
Apm – antenna pointing mechanism
Apollo – US manned lunar program.
Arc-second – 1/60 of an arc minute; 1/3600 of a complete rotation.
Arduino – a small, inexpensive microcontroller architecture.
Arinc – Aeronautical Radio, Inc. commercial company supporting

transportation, and providing standards for avionics.
ARM – Acorn RISC machine; a 32-bit architecture with wide application in embedded systems.
ARPA – (U. S.) Advanced Research Projects Agency.
ArpaNet – Advanced Research Projects Agency (U.S.), first packet switched network, 1968.
ASIN – Amazon Standard Inventory Number
Async – non synchronized
ATAC – Applied Technologies Advanced Computer.
AU – astronomical unit. Roughly 93 million miles, the mean distance between Earth and Sun,
BAE – British Aerospace.
Baud – symbol rate; may or may not be the same as bit rate.
Beowolf – a cluster of commodity computers; multiprocessor, using Linux.
Big-endian – data format with the most significant bit or byte at the lowest address, or transmitted first.
Binary – using base 2 arithmetic for number representation.
BIST – built-in self test.
Bit – binary variable, value of 1 or 0.
Boolean – a data type with two values; an operation on these data types; named after George Boole, mid-19th century inventor of Boolean algebra.
Bootloader – initial program run after power-on or reset. Gets the computer up & going.
Bootstrap – a startup or reset process that proceeds without external intervention.
BP - bundle protocol, for dealing with errors and disconnects.
BSP – board support package. Customization Software and device drivers.
Buffer – a temporary holding location for data.
Bug – an error in a program or device.
Bus – an electrical connection between 2 or more units; the engineering part of the spacecraft.
byte – a collection of 8 bits
C – programming language from Bell Labs, circa 1972.
CCSDS – Consultive Committee on Space Data Systems.

CDR – critical design review
C&DH – Command and Data Handling
Centaur – a dwarf planet.
cFE – Core Flight Executive – NASA GSFC reusable flight software.
CFS – Core Flight System – NASA GSFC reusable flight software.
Chip – integrated circuit component.
Clock – periodic timing signal to control and synchronize operations.
CME – Coronal Mass Ejection. Solar storm.
CogE – cognizant engineer for a particular discipline; go-to guy; specialist.
Complement – in binary logic, the opposite state.
Constellation – a grouping of satellites.
Control Flow – computer architecture involving directed flow through the program; data dependent paths are allowed.
COP – computer operating properly.
Coprocessor – another processor to supplement the operations of the main processor. Used for floating point, video, etc. Usually relies on the main processor for instruction fetch; and control.
Cots – commercial, off the shelf
CPU – central processing unit
CRC – cyclic redundancy code – error detection and correction mechanism.
Cubesat – small inexpensive satellite for colleges, high schools, and individuals.
D/A – digital to analog conversion.
DAC – digital to analog converter.
DARPA – Defense advanced research projects agency.
dc – direct current.
DCE – data communications equipment; interface to the network.
Deadly embrace – a deadlock situation in which 2 processes are each waiting for the other to finish.
Digital – using discrete values for representation of states or numbers.
Dnepr – Russian space launch system compatible with Cubesats.

DOD – (U. S.) Department of Defense.
DOE – (U. S.) Department of Energy.
DOF – degrees of freedom.
Downlink – from space to earth.
DSP – digital signal processing/processor.
DTE – data terminal equipment; communicates with the DCE to get to the network.
DTN – delay tolerant networks.
DUT – device under test.
ECC – error correcting code
EDAC – error detecting and correction circuitry.
EGSE – electrical ground support equipment
EIA – Electronics Industry Association.
ELV – expendable launch vehicle.
Embedded system – a computer systems with limited human interfaces and performing specific tasks. usually part of a larger system.
EMC – electromagnetic compatibility.
EMI – electromagnetic interference.
EOL – end of life.
EOS – Earth Observation spacecraft.
Ephemeris – orbital position data.
EPS – electrical power subsystem.
ESA – European Space Organization.
ESD – electrostatic discharge.
ESRO – European Space Research Organization
ESTO – NASA/GSFC – Earth Science Technology Office.
Ethernet – networking protocol, IEEE 802.3
ev – electron volt, unit of energy
EVA – extra-vehicular activity.
EXPRESS racks – on the ISS, EXpedite the PRocessing of Experiments for Space Station Racks
FAA – (U S.) Federal Aviation Administration.
Fail-safe – a system designed to do no harm in the event of failure.
Falcon – launch vehicle from SpaceX.
FCC – (U.S.) Federal Communications Commission.
FDC – fault detection and correction.

Femtosatellites - smaller than a Cubesat, 3.5 cm on a side.
Firewire – IEEE-1394 standard for serial communication.
Firmware – code contained in a non-volatile memory.
Fixed point – computer numeric format with a fixed number of digits or bits, and a fixed radix point. Integers.
Flag – a binary state variable.
Flash – non-volatile memory
Flatsat – prototyping and test setup, laid out on a bench for easy access.
Floating point – computer numeric format for real numbers; has significant digits and an exponent.
FPU – floating point unit, an ALU for floating point numbers.
Fram – ferromagnetic RAM; a non-volatile memory technology
FRR – Flight Readiness Review
FSW – flight software.
FTP – file transfer protocol
Gbyte – 10^9 bytes.
GEO – geosynchronous orbit.
GeV – billion (10^9) electron volts.
GMT – Greenwich Mean Time – a reference.
GNC – guidance, navigation, and control.
GPIO – general purpose I/O.
GPS – Global Positioning system – Navigation satellites.
GPU – graphics processing unit. ALU for graphics data.
GSFC – Goddard Space Flight Center, Greenbelt, MD.
Gyro – (gyroscope) a sensor to measure rotation.
Half-duplex – communications in two directions, but not simultaneously.
Handshake – co-ordination mechanism.
Hertz – cycles per second.
Hexadecimal – base 16 number representation.
Hi-rel – high reliability
HPCC – High Performance Computing and Communications.
IARU – International Amateur Radio Union
ICD – interface control document.
IC&DH – Instrument Command & Data Handling.
IEEE – Institute of Electrical and Electronic engineers

IEEE-754 – standard for floating point representation and calculation.
IMU – inertial measurement unit.
INPE Instituto Nacional de Pesquisas Espaciais (Brazilian National Institute for Space Research)
Integer – the natural numbers, zero, and the negatives of the natural numbers.
Interrupt – an asynchronous event to signal a need for attention (example: the phone rings).
Interrupt vector – entry in a table pointing to an interrupt service routine; indexed by interrupt number.
IP – intellectual property; Internet protocol.
IP-in-Space – Internet Protocol in Space.
IR – infrared, 1-400 terahertz. Perceived as heat.
IRAD – Independent Research & Development.
ISA – instruction set architecture, the software description of the computer.
ISBN – International Standard Book Number.
ISO – International Standards Organization.
ISR – interrupt service routine, a subroutine that handles a particular interrupt event.
ISS – International Space Station
I&T – integration & test
ITAR – International Trafficking in Arms Regulations (US Dept. of State)
ITU – International Telecommunications Union
IV&V – Independent validation and verification.
JEM – Japanese Experiment Module, on the ISS.
JHU – Johns Hopkins University.
JPL – Jet Propulsion Laboratory
JSC – Johnson Space Center, Houston, Texas.
JTAG – Joint Test Action Group; industry group that lead to IEEE 1149.1, Standard Test Access Port and Boundary-Scan Architecture.
JWST – James Webb Space Telescope – follow on to Hubble.
Kbps – kilo (10^3) bits per second.
Kernel – main portion of the operating system. Interface between

the applications and the hardware.
Kev – kilo electron volts. A measure of charge.
Kg – kilogram.
kHz – kilo (10^3) hertz
KVA – kilo volts amps – a measure of electrical power
Ku band – 12-18 Ghz radio
Lan – local area network, wired or wireless
LaRC – (NASA) Langley Research Center.
Latchup – condition in which a semiconductor device is stuck in one state.
Lbf – pounds-force.
LEO – low Earth orbit.
Let- Linear Energy Transfer
Lidar – optical radar.
Linux – open source operating system.
Logic operation – generally, negate, AND, OR, XOR, and their inverses.
LRR – launch readiness review
LRU – least recently used; an algorithm for item replacement in a cache.
LSB – least significant bit or byte.
LSP – (NASA) launch services program, or launch services provider
LUT – look up table.
Master-slave – control process with one element in charge. Master status may be exchanged among elements.
Mbps – mega (10^6) bits per second.
Mbyte – one million (10^6 or 2^{20}) bytes.
MEMS – Micro Electronic Mechanical System.
MET – mission elapsed time.
MEV – million electron volts.
MHz – one million (10^6) Hertz
Microcontroller – monolithic cpu + memory + I/O.
Microprocessor – monolithic cpu.
Microsat – satellite with a mass between 10 and 100 kg.
Microsecond – 10^{-6} second.
Microkernel – operating system which is not monolithic. So

functions execute in user space.
MLI – multi-layer insulation.
MPA – multiple payload adapter for deploying multiple p-pod's
MPE – Maximum predicted environments.
mram – magnetorestrictive random access memory.
mSec – Millisecond; (10^{-3}) second.
MIPS – millions of instructions per second.
MMU – memory management unit; manned maneuvering unit.
MSB – most significant bit or byte.
Multiplex – combining signals on a communication channel by sampling.
Multicore – multiple processing cores on one substrate or chip; need not be identical.
Mutex – a software mechanism to provide mutual exclusion between tasks.
Nano – 10^{-9}
NanoRacks – a company providing a facility onboard the ISS to support Cubesats
nanoSat – small satellite with a mass between 1 and 10 kg.
NASA - National Aeronautics and Space Administration.
NDA – non-disclosure agreement; legal agreement protecting IP.
NEN – (NASA's) Near Earth Network
Nibble – 4 bits, ½ byte.
NIST – National Institute of Standards and Technology (US), previously, National Bureau of Standards.
NMI – non-maskable interrupt; cannot be ignored by the software.
NRCSD - NanoRack CubeSat Deployer
NRE – non-recurring engineering; one-time costs for a project.
NSF – (U.S.) National Science Foundation.
NSR – non-space rated.
NTIA (U.S.) National Telecommunications and Information Administration
NUMA – non-uniform memory access for multiprocessors; local and global memory access protocol.
NVM – non-volatile memory.
NWS – (U.S.) National Weather Service
OBC – on board computer

OBD – On-Board diagnostics.
OBP – On Board Processor
Off-the-shelf – commercially available; not custom.
Open source – methodology for hardware or software development with free distribution and access.
Operating system – software that controls the allocation of resources in a computer.
Orbitel – orbiting hotel.
OSAL – operating system abstraction layer.
OSI – Open systems interconnect model for networking, from ISO.
Overflow - the result of an arithmetic operation exceeds the capacity of the destination.
Packet – a small container; a block of data on a network.
Paging – memory management technique using fixed size memory blocks.
Paradigm – a pattern or model
Paradigm shift – a change from one paradigm to another. Disruptive or evolutionary.
Parallel – multiple operations or communication proceeding simultaneously.
Parity – an error detecting mechanism involving an extra check bit in the word.
Pc – personal computer.
PCB – printed circuit board.
pci – personal computer interface (bus).
PCM – pulse code modulation.
PDR – preliminary design review
Peta - 10^{15} or 2^{50}
Phonesat – small satellite using a cell phone for onboard control and computation.
Picosat – small satellite with a mass between 0.1 and 1 kg.
Piezo – production of electricity by mechanical stress.
Pinout – mapping of signals to I/O pins of a device.
PiSat – a Cubesat architecture developed at NASA-GSFC, based on the Raspberry Pi architecture.
Pixel – picture element; smallest addressable element on a display or a sensor.

Plasma – an ionized state of matter
PLL – phase locked loop.
PocketQube – smaller than a Cubesat; 5 cm cubed, a mass of no more than 180 grams, and uses COTS components.
Poc – point of contact
POSIX – IEEE standard operating system.
PPF – payload processing facility
PPL – preferred parts list (NASA).
P-POD – Cubesat launch dispenser, Poly-Picosatellite Orbital Deployer
Psia – pounds per square inch, absolute.
PSP – Platform Support Package.
PWM – pulse width modulation.
Quadrature encoder – an incremental rotary encoder providing rotational position information.
Queue – first in, first out data buffer structure; implemented in hardware or software.
Rad – unit of radiation exposure
Rad750 – A radiation hardened IBM PowerPC cpu.
Radix point – separates integer and fractional parts of a real number.
RAID – redundant array of inexpensive disks.
Ram – random access memory.
RBF – remove before flight.
Real-time – system that responds to events in a predictable, bounded time.
Register – temporary storage location for a data item.
Reset – signal and process that returns the hardware to a known, defined state.
RF – radio frequency
RFC – request for comment
RISC – reduced instruction set computer.
RHPPC – Rad-Hard Power PC.
RHS – rad-hard software
RISC – reduced instruction set computer.
Router – networking component for packets.
RS-232/422/423 – asynchronous and synchronous serial

communication standards.
RT – remote terminal.
RTC – real time clock.
RTOS – real time operating system.
SAA – South Atlantic anomaly. High radiation zone..
Sandbox – an isolated and controlled environment to run untested or potentially malicious code.
SDR – software defined radio
Segmentation – dividing a network or memory into sections.
Semiconductor – material with electrical characteristics between conductors and insulators; basis of current technology processor, memory, and I/O devices, as well as sensors.
Semaphore – a binary signaling element among processes.
SD – secure digital (non-volatile memory card).
SDVF – Software Development and Validation Facility.
SEB – single event burnout.
SEL – single event latchup
Sensor – a device that converts a physical observable quantity or event to a signal.
Serial – bit by bit.
SEU – single event upset (radiation induced error).
Servo – a control device with feedback.
Six-pack – a six U Cubesat, 10 x 20 x 30 cm.
SN – (NASA's) Space Network
SOA – safe operating area; also, state of the art.
SOC – system on a chip; also state-of-charge.
Socket – an end-point in communication across a network
Soft core – a hardware description language description of a cpu core.
Software – set of instructions and data to tell a computer what to do.
Spacewire – high speed (160 Mbps) link.
SPI - Serial Peripheral Interface - a synchronous serial communication interface.
SRAM – static random access memory.
STAR – self test and repair.
STOL – system test oriented language, a scripting language for

testing systems.
Synchronous – using the same clock to coordinate operations.
Swarm – a collection of satellites that can operate cooperatively.
sync – synchronize, synchronized.
System – a collection of interacting elements and relationships with a specific behavior.
T&I – test and integration.
Terrabyte – 10^{12} bytes.
TCP/IP – Transmission Control Protocol/Internet protocol.
TDRSS – Tracking and Data Relay satellite system.
Tera - 10^{12} or 2^{40}
TCP/IP – transmission control protocol/internet protocol; layered set of protocols for networks.
TID – total ionizing dose.
TMR – triple modular redundancy.
Toolchain – set of software tools for development.
Transceiver – receiver and transmitter in one box.
Transducer – a device that converts one form of energy to another.
Train – a series of satellites in the same or similar orbits, providing sequential observations.
Triplicate – using three copies (of hardware, software, messaging, power supplies, etc.). for redundancy and error control.
Trojan – a body in the same orbit as a primary, but ahead or behind.
TRL – technology readiness level
Truncate – discard. cutoff, make shorter.
TT&C – tracking, telemetry, and command.
ttl – transistor-transistor logic integrated circuit.
UART – Universal asynchronous receiver-transmitter.
UDP – User datagram protocol; part of the Internet Protocol.
uM – micro (10^{-6}) meter
Underflow – the result of an arithmetic operation is smaller than the smallest representable number.
UoSat – a family of small spacecraft from Surrey Space Technology Ltd. (UK).
USAF – United States Air Force.
USB – universal serial bus.

VDC – volts, direct current.
Vector – single dimensional array of values.
VIA – vertical conducting pathway through an insulating layer.
Virtual memory – memory management technique using address translation.
Virtualization – creating a virtual resource from available physical resources.
Virus – malignant computer program.
Viterbi Decoder – a maximum likelihood decoder for data encoded with a Convolutional code for error control. Can be implemented in software or hardware
WiFi – short range digital radio.
Watchdog – hardware/software function to sanity check the hardware, software, and process; applies corrective action if a fault is detected; fail-safe mechanism.
Wiki – the Hawaiian word for "quick." Refers to a collaborative content website.
Word – a collection of bits of any size; does not have to be a power of two.
X-band – 7 – 11 GHz.
Zener – voltage reference diode.
Zero address – architecture using implicit addressing, like a stack.
Zombie-sat – a dead satellite, in orbit.
Zone of Exclusion – volume in which the presence of an object or personnel, or activities are prohibited.

If you enjoyed this book, you might also be interested in some of these.

Stakem, Patrick H. *16-bit Microprocessors, History and Architecture*, 2013 PRRB Publishing, ISBN-1520210922.

Stakem, Patrick H. *4- and 8-bit Microprocessors, Architecture and History*, 2013, PRRB Publishing, ISBN-152021572X,

Stakem, Patrick H. *Apollo's Computers,* 2014, PRRB Publishing, ISBN-1520215800.

Stakem, Patrick H. *The Architecture and Applications of the ARM Microprocessors,* 2013, PRRB Publishing, ISBN-1520215843.

Stakem, Patrick H. *Earth Rovers: for Exploration and Environmental Monitoring,* 2014, PRRB Publishing, ISBN-152021586X.

Stakem, Patrick H. *Embedded Computer Systems, Volume 1, Introduction and Architecture*, 2013, PRRB Publishing, ISBN-1520215959.

Stakem, Patrick H. *The History of Spacecraft Computers from the V-2 to the Space Station*, 2013, PRRB Publishing, ISBN-1520216181.

Stakem, Patrick H. *Floating Point Computation*, 2013, PRRB Publishing, ISBN-152021619X.

Stakem, Patrick H. *Architecture of Massively Parallel Microprocessor Systems*, 2011, PRRB Publishing, ISBN-1520250061.

Stakem, Patrick H. *Multicore Computer Architecture,* 2014, PRRB

Publishing, ISBN-1520241372.

Stakem, Patrick H. *Personal Robots*, 2014, PRRB Publishing, ISBN-1520216254.

Stakem, Patrick H. *RISC Microprocessors, History and Overview,* 2013, PRRB Publishing, ISBN-1520216289.

Stakem, Patrick H. *Robots and Telerobots in Space Applications*, 2011, PRRB Publishing, ISBN-1520210361.

Stakem, Patrick H. *The Saturn Rocket and the Pegasus Missions, 1965,* 2013, PRRB Publishing, ISBN-1520209916.

Stakem, Patrick H. *Visiting the NASA Centers, and Locations of Historic Rockets & Spacecraft,* 2017, PRRB Publishing, ISBN-1549651205.

Stakem, Patrick H. *Microprocessors in Space*, 2011, PRRB Publishing, ISBN-1520216343.

Stakem, Patrick H. Computer *Virtualization and the Cloud*, 2013, PRRB Publishing, ISBN-152021636X.

Stakem, Patrick H. *What's the Worst That Could Happen? Bad Assumptions, Ignorance, Failures and Screw-ups in Engineering Projects, 2014,* PRRB Publishing, ISBN-1520207166.

Stakem, Patrick H. *Computer Architecture & Programming of the Intel x86 Family, 2013,* PRRB Publishing, ISBN-1520263724.

Stakem, Patrick H. *The Hardware and Software Architecture of the Transputer,* 2011,PRRB Publishing, ISBN-152020681X.

Stakem, Patrick H. *Mainframes, Computing on Big Iron*, 2015, PRRB Publishing, ISBN- 1520216459.

Stakem, Patrick H. *Spacecraft Control Centers*, 2015, PRRB Publishing, ISBN-1520200617.

Stakem, Patrick H. *Embedded in Space,* 2015, PRRB Publishing, ISBN-1520215916.

Stakem, Patrick H. *A Practitioner's Guide to RISC Microprocessor Architecture*, Wiley-Interscience, 1996, ISBN 0471130184.

Stakem, Patrick H. *Cubesat Engineeering*, PRRB Publishing, 2017, ISBN-1520754019.

Stakem, Patrick H. *Cubesat Operations*, PRRB Publishing, 2017, ISBN-152076717X.

Stakem, Patrick H. *Interplanetary Cubesats*, PRRB Publishing, 2017, ISBN-1520766173 .

Stakem, Patrick H. Cubesat Constellations, Clusters, and Swarms, Stakem, PRRB Publishing, 2017, ISBN-1520767544.

Stakem, Patrick H. *Graphics Processing Units, an overview*, 2017, PRRB Publishing, ISBN-1520879695.

Stakem, Patrick H. *Intel Embedded and the Arduino-101, 2017,* PRRB Publishing, ISBN-1520879296.

Stakem, Patrick H. *Orbital Debris, the problem and the mitigation*, 2018, PRRB Publishing, ISBN-*1980466483*.

Stakem, Patrick H. *Manufacturing in Space*, 2018, PRRB Publishing, ISBN-1977076041.

Stakem, Patrick H. *NASA's Ships and Planes*, 2018, PRRB Publishing, ISBN-1977076823.

Stakem, Patrick H. *Space Tourism*, 2018, PRRB Publishing, ISBN-

1977073506.

Stakem, Patrick H. *STEM – Data Storage and Communications*, 2018, PRRB Publishing, ISBN-1977073115.

Stakem, Patrick H. *In-Space Robotic Repair and Servicing*, 2018, PRRB Publishing, ISBN-1980478236.

Stakem, Patrick H. *Introducing Weather in the pre-K to 12 Curricula, A Resource Guide for Educators*, 2017, PRRB Publishing, ISBN-1980638241.

Stakem, Patrick H. *Introducing Astronomy in the pre-K to 12 Curricula, A Resource Guide for Educators*, 2017, PRRB Publishing, ISBN-198104065X.
Also available in a Brazilian Portuguese edition, ISBN-1983106127.

Stakem, Patrick H. *Deep Space Gateways, the Moon and Beyond*, 2017, PRRB Publishing, ISBN-1973465701.

Stakem, Patrick H. *Exploration of the Gas Giants, Space Missions to Jupiter, Saturn, Uranus, and Neptune*, PRRB Publishing, 2018, ISBN-9781717814500.

Stakem, Patrick H. *Crewed Spacecraft*, 2017, PRRB Publishing, ISBN-1549992406.

Stakem, Patrick H. *Rocketplanes to Space*, 2017, PRRB Publishing, ISBN-1549992589.

Stakem, Patrick H. *Crewed Space Stations*, 2017, PRRB Publishing, ISBN-1549992228.

Stakem, Patrick H. *Enviro-bots for STEM: Using Robotics in the pre-K to 12 Curricula, A Resource Guide for Educators*, 2017, PRRB Publishing, ISBN-1549656619.

Stakem, Patrick H. *STEM-Sat, Using Cubesats in the pre-K to 12 Curricula, A Resource Guide for Educators*, 2017, ISBN-1549656376.

Stakem, Patrick H. *Lunar Orbital Platform-Gateway*, 2018, PRRB Publishing, ISBN-1980498628.

Stakem, Patrick H. Embedded GPU's, 2018, PRRB Publishing, ISBN- 1980476497.

Stakem, Patrick H. RISC-, and Open Source Solution for Space Flight Computers, 2019, PRRB Publishing, ISBN-1796434388.

Stakem, Patrick H. Mobile Cloud Robotics, 2018, PRRB Publishing, ISBN- 1980488088

Stakem, Patrick H. *Extreme Environment Embedded Systems*, 2017, PRRB Publishing, ISBN-1520215967.

Stakem, Patrick H. *What's the Worst, Volume-2*, 2018, ISBN-1981005579.

Stakem, Patrick H., *Spaceports*, 2018, ISBN-1981022287.

Stakem, Patrick H., *Space Launch Vehicles*, 2018, ISBN-1983071773.

Stakem, Patrick H. *Mars*, 2018, ISBN-1983116902.

Stakem, Patrick H. *X-86, 40th Anniversary ed*, 2018, ISBN-1983189405.

Stakem, Patrick H. *Lunar Orbital Platform-Gateway*, 2018, PRRB Publishing, ISBN-1980498628.

Stakem, Patrick H. *Space Weather*, 2018, ISBN-1723904023.

Stakem, Patrick H. *STEM-Engineering Process*, 2017, ISBN-1983196517.

Stakem, Patrick H. *Space Telescopes,* 2018, PRRB Publishing, ISBN-1728728568.

Stakem, Patrick H. *Exoplanets*, 2018, PRRB Publishing, ISBN-9781731385055.

Stakem, Patrick H. *Planetary Defense*, 2018, PRRB Publishing, ISBN-9781731001207.

Patrick H. Stakem *Exploration of the Asteroid Belt*, 2018, PRRB Publishing, ISBN-1731049846.

Patrick H. Stakem *Terraforming*, 2018, PRRB Publishing, ISBN-1790308100.

Patrick H. Stakem, *Martian Railroad,* 2019, PRRB Publishing, ISBN-1794488243.

Patrick H. Stakem, *Exoplanets,* 2019, PRRB Publishing, ISBN-1731385056.

Patrick H. Stakem, *Exploiting the Moon,* 2019, PRRB Publishing, ISBN-1091057850.

Patrick H. Stakem, *RISC-V, an Open Source Solution for Space Flight Computers,* 2019, PRRB Publishing, ISBN-1796434388.

www.ingramcontent.com/pod-product-compliance
Lightning Source LLC
Chambersburg PA
CBHW030451220526

45464CB00006B/2489